Lecture Notes in Physics

Edited by J. Ehlers, München, K. Hepp, Zürich
R. Kippenhahn, München, H. A. Weidenmüller, Heidelberg
and J. Zittartz, Köln
Managing Editor: W. Beiglböck, Heidelberg

122

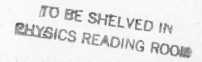
New Developments in Semiconductor Physics

Proceedings of the International Summer School
Held in Szeged, Hungary July 1 – 6, 1979

Edited by
F. Beleznay, G. Ferenczi, and J. Giber

Springer-Verlag
Berlin Heidelberg New York 1980

Editors

Dr. Ferenc Beleznay
Dr. György Ferenczi
Research Institute for
Technical Physics of the
Hungarian Academy of Sciences
1325 Budapest, Pf. 76
Hungary

Prof. Dr. János Giber
Institute of Physics of the
Technical University of Budapest
1111 Budapest, Budafoki ut 8
Hungary

ISBN 3-540-09988-3 Springer-Verlag Berlin Heidelberg New York
ISBN 0-387-09988-3 Springer-Verlag New York Heidelberg Berlin

© by Springer-Verlag Berlin Heidelberg 1980
Printed in Germany

Printing and binding: Beltz Offsetdruck, Hemsbach/Bergstr.
2153/3140-543210

PREFACE

This volume consists of lecture notes and selected contributed papers presented at the International Summer School on New Developments in Semiconductor Physics held at the University of Szeged, July 1-6, 1979. The major part of the contributions in this volume is related to the new experimental technics and theoretical ideas applied in research of new semiconductor materials, mostly III-V semiconductors.

We wish to thank the staff of the Department of Physics in the University of Szeged and of the Institute of Physics of the Technical University of Budapest for their cooperation in organizing this school and preparing this volume and Zsuzsa Nagy for her patient job of typing the manuscript.

Budapest, January 1980,

F. Beleznay   J. Giber   G. Ferenczi

TABLE OF CONTENTS

LOW TEMPERATURE PHOTO- AND MAGNETO-TRANSPORT
INVOLVING IMPURITY-PHONON RESONANCES IN SEMICONDUCTORS

R.A. Stradling
Physics Department
University of St Andrews
Fife KY16 9SS, U.K.

ABSTRACT

This paper reviews a number of resonance effects involving impurities and phonons. In a variant of the magnetophonon effect, electrons can loose energy from Landau levels by emitting L.O. phonons and being captured at impurities. This effect is demonstrated for n-Si and the intervalley phonons involved are determined. In the magneto-impurity effect inelastic scattering at the impurity takes the role of the optic phonons in the magneto-phonon effect. This effect and the impurity capture effect are found with n-CdTe. With this material capture at the n=1 and n=2 donor states is responsible for the series of deps   in the photoresponse known   as "oscillatory photoconductivity". Oscillatory photoconductivity involving the shallow donors is also demonstrated for n-CdS. Oscillatory photoconductivity involving a number of excited states of several of the shallow donors has recently been reported for n-GaP.

1. INTRODUCTION

The first direct evidence that carriers could lose energy by emitting L.O. phonons while being captured at impurity sites came from the observation of a periodic variation of the photoresponse of diamond with the energy of the photon used to generate the photosignal (Hardy et.al. 1962). It was realised at that time the photocurrent would be quenched if the carriers found themselves at the correct energy to be captured at impurity states with the emission of one or more L.O. phonons. The period observed corresponded to the energy of the L.O. phonons ($\hbar\omega_{LO}$) and minima in the photoresponse were observed at photon frequencies (ω) corresponding to

$$\hbar\omega = N\hbar\omega_{LO} + \Delta E_I \qquad (1)$$

where ΔE_I is the difference in energy between the ground state of the impurity and the excited state involved. Stocker (1966) considered the dynamics of the process where ΔE_I equalled the difference between the ground state of the impurity and the nearest band edge (i.e. the impurity binding energy E_I). While studying the oscillatory photoresponse of CdTe Mears et al. (1968) were the first to detect the involvement of the impurity ground state and therefore observed a process where ΔE_I is zero in equation (1). The application of a magnetic field introduces the possibility of additional resonances as the band states become quantised into Landau levels. In this case further singularities in the photoresponse are expected when the carriers after photoexcitation from the impurity states and subsequent photoemission of N optical phonons find themselves in the mth Landau state. When this happens a minimum in the photoresponse occurs when

$$\hbar \omega = N \hbar \omega_{LO} + m \hbar \omega_c + E_I(B) \tag{2}$$

where ω_c is the cyclotron frequency and E_I is now dependent on the magnetic field B. A better known resonance effect which occurs without photoexcitation is the magnetophonon or Gurevich-Firshov effect (Gurevich et al 1963) which occurs when the carrier is scattered between them and the m + nth Landalu level with the emission or absorption of L.O. phonons. In this case extrema are observed in the magnetoresistance at magnetic fields given by

$$\hbar \omega_{LO} = n \hbar \omega_c = n \hbar eB_{n/m^*} \tag{3}$$

where B_n are the resonance fields and m^* is the effective mass of the band states concerned. In 1970 it was realised by Stradling & Wood that capture at impurity states was alternative process to the magnetophonon effect which was favoured at low temperatures. In this case peaks in the magnetoresistance could be observed at fields given by

$$\hbar \omega_{LO} = n \hbar \omega_c + E_I(B) \tag{4}$$

Transport processes involving resonances described by equations (1)-(4) have been reviewed by Harper et al (1973). A final resonance effect which can be observed in transport experiments is the magneto-impurity effect where an impurity is resonantly excited or de-excited by scattering carriers between Landau states so that

$$\Delta E_I(B) = n\hbar\omega_c \tag{5}$$

In equation (5) ΔE_I can either be the energy between the ground and
excited states of the impurity or the binding energy of the impurity.
The magneto-impurity effect was first observed with n-InP (Eaves et al
1974) and has subsequently been reviewed by Eaves and Portal (1978,
1979). The present paper discusses some recent developments involving
impurity-L.O. phonon resonances in semiconductors.

2. IMPURITY ASSOCIATED MAGNETOPHONON RESONANCE IN n-TYPE SILICON

Recently the first observation of magneto-phonon resonances associated with impurity capture in n-type silicon has been reported
(Portal et al 1979). The experiments were performed at lower temperatures (25-40 K) than those employed in earlier magnetophonon studies of the intervalley scattering in silicon (50-70 K) (Portal et al 1974, Eaves et
al 1974). This is the first observation of such a process in a multi-
valley semiconductor although the involvement of impurities in hot
electron magnetophonon resonance has been well established in materials
with conduction bands located at the centre of the Brillouin zone.
(See review by Harper et al 1973). The additional complication in the
present experiments is that the impurity capture processes involve intervalley phonons. Furthermore, in contrast to previous observations
of the magneto-phonon effect associated with capture at impurities,
(Harper et al 1973) the binding energies of the shallow donors are
comparable to the phonon energies, and the binding energies are multi-
valued due to the valley-orbit splitting of the ground state. In
addition the several different phonons which are capable of relaxing
electron energy by intervalley scattering, combine with the different
1s donor states to give rise to many possible energy relaxation
mechanisms. The energies of these resonant relaxation mechanisms are
shifted to values which are typically between 2 and 10 times lower
than the original phonon energies. This shift has enabled a more accurate determination of the phonon energies involved, together with
the observation of additional phonons not previously detected in
earlier magnetophonon experiments with silicon.

The influence of uniaxial compressive stress has been studied, on
both the series fundamental fields and the amplitudes of the oscillatory
structure. The presence of uniaxial stress alters the resonance conditions through the change in the relative energies of the different

conduction band minima. At large values of applied uniaxial stress all
f-scattering processes, which occur between non-equivalent valleys,
are completely suppressed leaving only g-scattering between the two
conduction band mimina parallel to the applied stress. In addition the
reduction in degeneracy of the conduction band minima changes the
binding energies of the different 1s donor states.

IMPURITY CAPTURE RESONANCE CONDITIONS

Magnetophonon oscillations were first observed in n-type silicon
by Portal et al (1974). They observed resonances from electrons
heated by the electric field out of equilibrium with the lattice, which
was maintained at a fixed temperature between 55 K and 77 K. Six different phonon scattering processes were found to contribute to the
energy relaxation of the electrons.

The multi-ellipsoidal structure of the conduction band in silicon
leads to the possibility of intervalley scattering by either acoustic
or optic phonons which may be between ellipsoids with parallel principal axes (g-scattering) or between ellipsoids situated along perpendicular crystals axes (f-scattering). In addition, the very aniso-
tropic cyclotron masses produce a magnetophonon spectrum which varies
markedly with the direction of the applied magnetic field. The simplest
case occurs when B $\parallel$ E $\parallel$ (111) and all six valleys are equivalent.
Magnetophonon resonances will be determined for both f- and g-scattering by the relation

$$\hbar\omega_i = \frac{Ne\hbar B}{m_{111}^*}$$

(1)

where m_{111}^* is the cyclotron mass for B $\parallel$ (111).

With B $\parallel$ (100) and B $\parallel$ (110) the resonance conditions are more complex due to the non-equivalence of the six valleys. The condition for
resonant f-scattering is

$$\pm\hbar\omega_i = (N + 1/2)\, \frac{e\hbar B}{m_1} - (m + 1/2)\, \frac{e\hbar B}{m_2}$$

(2)

for scattering between valleys with different cyclotron masses m_1 and
m_2 and

$$\hbar\omega_i = \frac{Ne\hbar B}{m_2}$$

(3)

for valleys with the same cyclotron mass.

Equation (2) describes energy relaxation from either transverse to longitudinal valleys or from longitudinal to transverse valleys and each of these processes will give rise to a set of magnetophonon series with differing phase. Thus there will be two periodicities produced in the Fourier analysis which are given by the relations $\Delta(^1/B_1) = ^e/m_1\omega_o$ (M = const.) and $\Delta(^1/B_2 = ^e/m_2\omega_o$ (N = const.). For g-scattering the resonance condition is

$$\hbar\omega_i = \frac{Ne\hbar B}{m_{1,2}} \qquad (4)$$

For $B \parallel (100)$ there are two longitudinal valleys with an effective mass $m_1 = m_\perp$ and there are four transverse valleys with a mass $m_2 = \sqrt{m_\perp m_\parallel}$. In the case of $B \parallel (110)$ there are four longitudinal valleys with an effective mass given by

$$^1/m_1 = \frac{1}{\sqrt{2}} \left(\frac{1}{m_\perp^2} + \frac{1}{m_\perp m_\parallel}\right)^{1/2} \qquad (5)$$

and there are two transverse valleys with a mass $m_2 = \sqrt{m_\perp m_\parallel}$. In the analysis of the experimental data of the present paper the transverse cyclotron mass m_t has been increased slightly from its low temperature value of $0.1905\, m_o$ to values between $0.195\, m_o$ (160K) and $0.198\, m_o$ (340K) to account for non-parabolicity within the conduction band (Stradling and Zhukov 1966, Ousset et al 1976) in the same way as was done by Eaves et al (1975) in order to obtain a good fit of their magnetophonon data. Similarly the value taken for m_l ($0.90\, m_o$) is slightly smaller than the band edge value of $0.916\, m_o$ measured by Hensel and co-workers (Hensel et al 1965) but is more consistent with the observed anisotropy of the magnetophonon series.

The experiments of Portal et al (1974) and Eaves et.al (1974, 1975) have shown that phonons of energies 140K(g), 220K(g or f), 530K(f), 585K(f), 685K(f) and 745K(g) all give rise to the magneto-phonon series described by equations (1) to (5).

Magnetophonon series associated with warm-electron energy relaxation frequently involve capture at the ground states of shallow donor impurities, giving rise to oscillations determined by the resonance condition

$$\hbar\omega_i = N\hbar\omega_c + E_I(B) \qquad (6)$$

This mechanism was first demonstrated for GaAs (Stradling and Wood 1968) and has subsequently been verified for a number of semiconductors having conduction band edges located at the centre of the Brillouin zone (Harper et al 1973).

As was shown by Nicholas and Stradling (1976) the amplitude of impurity-capture associated resonances is strongly temperature dependent. Accordingly the magnetophonon studies of Portal et al (1974) in n-type silicon have been extended to lower temperatures ($\sim$ 30 K) where the electrons just begin to freeze out. In this temperature region several magnetophonon series associated with electron capture at impurity sites have been observed. This process is qualitatively different in silicon as compared with that found in direct gap semiconductors (Harper et al 1973), where the resonances occur from only one phonon, and the shallow donor energies are small compared with the phonon energy and are well described by the effective mass theory. In addition to complications in n-type silicon arising from the several different phonons which contribute to scattering for an arbirary direction of field, the effective mass theory for group V donors is inadequate when applied to the 1s ground state (Kohn and Luttinger 1955). The sixfold degeneracy for the states, originating from the six conducton band minima, is lifted in the case of the 1s state, which splits into a singlet $1s(A_1)$, a doublet $1s(E)$, and a triplet $1s(T_1)$.

The nature of all three 1s states is such that they contain contributions from all six conduction band minima. It is thus clear that impurity-capture associated magnetophonon series may occur in n-type silicon for all three 1s states for both f- and g-type scattering processes and that these will both be determined by the relation

$$\hbar\omega_i = \frac{N\,e\,\hbar\,B}{m_{1,2}} + E_I\,(B) \tag{7}$$

The following approximation for $E_I(B)$ may be used

$$E_I(B) = E_I(0) + \frac{1}{2}\frac{e\,\hbar\,B}{m_{1,2}} \tag{8}$$

provided that $\gamma = \frac{\hbar\omega_t}{2E_I(0)} < 1$ where $E_I(0)$ is the effective mass binding energy = 375°K. In the case of n-type silicon $\gamma = 0.1$ at 10 T for m_t so that the oscillations will be accurately determined by the relation

$$\hbar\omega_i - E_I(0) = (N + \frac{1}{2})\frac{e\,\hbar\,B}{m_{1,2}} \tag{9}$$

The Fourier analysis of such structure has been considered by Nicholas and Stradling (1976) and will give rise to a fundamental field given by

$$B_F = \left[\Delta(^1/B) \right]^{-1} = \frac{m_{1,2}}{e} (\hbar\omega_i - E_I(0)) \qquad (10)$$

The binding energies of the donor 1s states are comparable to the phonon energies and hence give rise to resonance energies which are between 5 and 10 times lower than for the normal phonon emission series. The possible resonance energies $\hbar\omega_i - E_I(0)$ are given in Table 1 for the various phonons and 1s states present.

Table 1. The possible resonance energies $(\hbar\omega_i - E_I(0))$ for impurity capture associated magnetophonon resonances in phosphorus doped silicon

Phosphorus impurity ground state energies	Intervalley phonon energies			
	735K(g)	685K(f)	585K(f)	535K(f)
1s(A) 527K	208K	158K	58K	-
1s(T) 392K	343K	293K	193K	143K
1s(E) 376K	359K	309K	209K	159K

It may be noted that the values quoted for the phonon energies differ slightly (at the most by 1 1/2 %) from those of Eaves et.al. (1975). The values have been adjusted to give a better fit to the impurity-capture magnetophonon series which are observed. The values of the high energy f and g phonons are thus concluded to be 735K(g), 685(f) 585K(f) and 535K(f). This should be a more accurate determination of the phonon energies, since the resonance energies are much smaller and there is less uncertainty in the non-parabolicity correction to the effective mass. Furthermore the binding energy which must then be added to the resonance energy to give the phonon energy is a precise value, which has been spectroscopically determined (Aggarwal and Ramdas 1965).

EXPERIMENTAL RESULTS AT ZERO STRESS

At low electric field for B//(100) and B//(110) the structure is totally dominated by the presence of one series of minima a $\frac{\partial^2\sigma}{\partial B^2}$ and hence maxima in σ, with a fundamental field of 68 Teslas. This series was observed by Eaves et al (1975) as a weak component among several others. It was interpreted in terms of a g or f phonon energy 220 K, and the heavy mass of 0.43 m_o. The measurements to be presented below with applied uniaxial stress show that this is an f-scattering process. As the electric field is raised the oscillations become more complex and Fourier analysis must be employed to determine which series are present. Several fundamentals appear at higher electric fields in each of the orientations, while at high electric fields the 68T series completely disappears. The most prominent features of the spectra are series with fundamentals of 23.5, 30.5 and 34.5 T for the three orientations respectively, giving a resonance energy of 160 K. These series are thought to arise from the emission of a 685K f-scattering phonon accompanied by capture at a 1s(A) ground state. The association of these series with impurity site capture is clearly demonstrated by plotting the peak number n versus 1/B. The half-integer intercept at 1/B = 0 is found which is characteristic of capture at the impurity sites as may be seen from inspection of the resonance conditions, equation (10). In addition to this strong series of oscillations there are several other series present, as predicted by equations (1) to (5). The observed fundamentals are given in Table 2 for all three orientations, together with the possible interpretation of the processes which give rise to the resonance energies. As may be seen there is a considerable uncertainity as to which processes give rise to the various fundamentals owing to the near coincidence of many of the resonance energies. Positive identification can be made in many cases by the measurements of the series fundamentals as a function of stress to be reported below.

THE INFLUENCE OF APPLIED UNIAXIAL STRESS

The previous section has shown that the capture of electrons at impurity sites accompanied by the emission of f- or g-scattering phonons may give rise to magnetophonon structure. Due to the large number of possible processes resulting from the presence of several different phonons and 1s-ground state energies, it was not possible to identify unambiguously which processes were giving rise to resonant structure in the magnetoresistance. In order to improve the understanding of these processes a study has been performed of the independences of the oscillations upon applied uniaxial stress. The particular usefulness of this technique was first demonstrated by Eaves et al (1974) who used it to distinguish between f- and g-type scattering in n-silicon at 60 K.

Table 2. The experimentally observed magnetophonon series fundamentals at zero stress

Orientation	Series fundamental field (Teslas)	m^*	Resonance energy (K)	Calculated energy form eq.(11) (K)	Capture process with phonon and final state
B ∥ (100)	23.5	m_1	164	157	685(f) − 1s(A)
				142	535(f) − 1s(T)
				157	535(f) − 1s(E)
	31.8	m_1	220	207	735(g) − 1s(A)
				192	585(f) − 1s(T)
				207	585(f) − 1s(E)
	40	m_1	275	292	685(f) − 1s(T)
				307	685(f) − 1s(E)
	49	m_2	158	157	685(f) − 1s(A)
				142	535(f) − 1s(A)
				157	535(f) − 1s(E)
	62	m_1	333	342	735(g) − 1s(T)
		m_2	200	207	735(g) − 1s(A)
				192	585(f) − 1s(T)
				207	585(f) − 1s(E)
	93	m_2	299	292	685(f) − 1s(T)
				307	685(f) − 1s(E)
B ∥ (110)	30.5	m_1	166	157	635(f) − 1s(A)
				142	535(f) − 1s(T)
				157	535(f) − 1s(E)
	64	m_1	206	207	735(g) − 1s(A)
				192	585(f) − 1s(T)
				207	585(f) − 1s(E)
B ∥ (111)	34.5		153	157	685(f) − 1s(A)
				142	535(f) − 1s(T)
				157	535(f) − 1s(E)
	42		186	207	735(g) − 1s(A)
				192	585(f) − 1s(T)
				207	585(f) − 1s(E)
	61		265	292	685(f) − 1s(T)
	67		295	307	685(f) − 1s(E)

(A simplified interpretation of the above impurity capture series is given in Table 4 in the light of the stress dependent results.)

In addition the following inter Landau level series, which result from the 220K f-scattering phonon, were observed at low electric fields:

B	(100)	68.0	m_2	219	220(f)
B	(110)	68.0	m_2	219	220(f)
B	(111)	48.5		216	220(f)

The values of effective mass m_1 and m_2 for B//(100) and B//(110) are given in the text

The six conduction band minima is lifted by an amount ΔE_s given by

$$\Delta E_s = \Sigma_u (S_{11} - S_{12}) X \qquad (11)$$

where Σ_u is the shear deformation potential constant and S_{11} and S_{12} are elastic compliance constants. Using the value for Σ_u of 8.9 eV obtained from measurements of the piezo-spectroscopy of shallow donors in silicon (Tekippe et al 1972), equation (11) yields a temperature equivalent energy shift of 98 K/kbar. A compressive stress of a few kilobars applied along a (100) direction almost totally depopulates the four valleys perpendicular to the (100) direction at a lattice temperature of 30 K, and suppresses f-scattering.

In addition the reduction of the conduction band degeneracy results in shifts and splittings of the 1s donor levels. The energies and compositions of the states have been calculated by Wilson and Feher (1961), and measured as a function of uniaxial stress up the high stress limit by Cooke et al (1978).

According to stress dependent cyclotron resonance (Hensel et al 1965), Raman scattering (Anastassakis et al 1970) and ultrasonic experiments (Hall 1967), the changes in cyclotron mass and phonon energy with increasing stress are very small and hence have been neglected.

The changes in energy levels will influence both the positions and amplitudes of the magnetophonon oscillations. Consideration of the condition for inter-Landau levels f-scattering gives

$$\pm \hbar \omega_i = (N + \tfrac{1}{2}) \frac{\hbar eB}{M_1 (100)} - \left\{ (M + \tfrac{1}{2}) \frac{\hbar eB}{M_2 (100)} + \Delta E_s \right\} \qquad (12)$$

$$N, M = 0, 1, 2, 3$$

so that there are now three possible resonant series. Firstly the resonance condition for f-scattering between valleys with principle axes perpendicular to (100) is

$$\hbar \omega_i = \frac{N \hbar eB}{M_2 (100)} \qquad (13)$$

and is unaffected by stress X//(100). Secondly for f-scattering from the transverse to the longitudinal valleys

$$\hbar \omega_i - \Delta E_s = (M + c) \frac{\hbar eB}{M_2 (100)} \qquad (14)$$

where c is a phase factor. Finally for f-scattering from the longitudinal to the transverse valleys the extrema are determined by the relation

$$\hbar\omega_i + \Delta E_s = (N + c')\frac{\hbar eB}{M_1}_{(100)}$$

(15)

All of these processes are inhibited by the shift and depopulation of the (001) and (010) valleys produced by the stress. The resonance conditions for inter-Landau level g-scattering are not affected by the stress.

The shift in the energy levels of the shallow donors will alter the resonance condition, equation (9), for electron capture at impurity sites. The resonance condition may now be written in the form

$$\hbar\omega_i - E_I(0, X)_{1,2} = (N + \frac{1}{2})\frac{e\hbar B}{M_{1,2}}$$

(16)

where the binding energy $E_I(0,X)_{1,2}$ is defined relative to the valleys from which the captured electrons originate. The values of $E_I(0,X)$ were taken from Wilson and Feher (1961) and the experimentally deter-mined values of Cooke et al (1978).

The splitting of the 1s energy levels by the stress results in the recomposition of some of the states, and the occurrence of others which do not contain contributions from all six valleys. It is now necessary to consider which capture processes are allowed by wavevec-tor conservation. In the case of the 1s(A_1) state ($\alpha^{\pm}_{11,21}$) certain processes become prohibited only at high stresses as a result of the change in the composition of the states. This may be illustrated by consideration of the processes which involve transitions from the lon-gitudinal valleys to the lowest ($\alpha^-_{11,21}$) 1s state, accompanied by f-phonon emission. Such a process gives rise to the dominant 160 K energy relaxation mechanism observed for all three orientations at zero stress. The amplitude of the oscillations will be proportional to the admixture of the transverse valleys present, and it may be calcu-lated from the relations given by Wilson and Feher (1961) that it is necessary to apply stresses above 2 kbar to strongly inhibit such f-scattering processes. The allowed and forbidden processes, as deduced from the composition of the different 1s states, are given in Table 3.

It may thus be concluded that the influence of uniaxial stress is to strongly suppress f-scattering processes and to rapidly depopulate the transverse valleys, thus suppressing all series with the large ef-fective mass $m_2(100)$. The depopulation of the transverse valleys is extremely rapid at 30 K where 0.2 kbar will reduce the relative popu-lation by a factor of two. Inter-Landau level f-scattering between low mass valleys is suppressed due to the rising electron energy required

Table 3. The phonon-impurity capture processes which are allowed and forbidden by momentum conservation in the presence of applied uniaxial stress.

		735(g)	685(f)	phonons (K) 585(f)	535(f)
(i) Capture from Low Mass (Longitudinal) Valleys					
1s(A)	(A$_1$)	✓	✓+	✓+	✓+
1s(T)	(B$_2$)	✓	X	X	X
	(E)	X	✓	✓	✓
1s(E)	(A$_1$)	✓+	✓	✓	✓
	(B$_1$)	X	✓	✓	✓
(ii) Capture from High Mass (Transverse) Valleys					
1s(A)	(A$_1$)	✓	✓	✓	✓
1s(T)	(B$_2$)	X	✓	✓	✓
	(E)	✓	✓	✓	✓
1s(E)	(A$_1$)	✓+	✓	✓	✓
	(B$_1$)	X	✓	✓	✓

Processes marked ✓ are allowed, while those marked X are forbidden and those marked + are fordibben in the high stress limit.

for a transition into the transverse valleys. Eaves et al (1974) have shown that at 60 K a stress of 1 kbar will remove such f-scattering processes. Impurity capture associated f-scattering processes are suppressed less strongly than this as the admixture of transverse valleys falls less rapidly. This is due to the large energy splitting (150 K) between the interacting states, which only becomes comparable to the stress splitting of the longitudinal and transverse valleys at ∼ 1.5 kbars.

5. EXPERIMENTAL RESULTS IN THE PRESENCE OF UNIAXIAL STRESS

The dependences of both the series fundamentals and the oscillatory amplitudes of the magnetophonon oscillations have been studied as a function of uniaxial stress up to 6 kbars applied in the (100) direction. At low electric fields the application of even 0.25 kbar is sufficient to produce oscillations from the series at 20.4 Teslas corresponding to the 140 K g-scattering phonon. The other series present start to change position and fall in amplitude with increasing stress so that at stresses above 2 kbars the only structure left corresponds to the 20.4 Teslas series, some second and third harmonics of this series, and a weak peak which occurs at 28.5 Teslas. The weak structure is only observable at stresses above 2 kbars and shows no dependence of position or amplitude upon stress. This fundamental of 28.5 Teslas corresponds to an energy of 200 K, and is probably due to g-scattering by a L.A., Δ_1 symmetry phonon whose energy was measured by Dolling (1963) and Palevsky et al (1959) as 220 K.

The group theoretical treatments of both Streitwolf (1970) and Lax and Birman (1972) show both the 140 K T.A. Δ_5 g-phonon and the 200 K L.A. Δ_1 g-phonons are forbidden in zero magnetic field (as is the 220 K T. Δ_4 f-scattering phonon). More recent calculations by Wallace and Joos (1978) show that the influence of the magnetic field and other perturbations without the lattice symmetry, such as impurities, is to relax many of the prohibitions. In particular the 140 K T.A. g-scattering and 220 K T.A. f-scattering phonons become quite strongly allowed, however the 220 K L.A. g-scattering phonon still remains forbidden. These results explain quite well the dominance of the 140 K g-phonon series at high stresses where all the allowed f-scattering processes are strongly suppressed. The mechanism for the weak violation of the selection rules for the 200 K L.A. g-phonon is not clear.

At higher electric fields the behaviour as a function of stress is at first similar to that at low electric fields as the application of weak stresses causes the appearance of the 20.4 Teslas series. At intermediate stresses there are several series present between 25 and 50 Teslas which show a weak dependence of series fundamentals upon stress and which gradually fall off in amplitude. At 3 kbars and above the spectra are dominated by the series at 20.4 Teslas and a group of two or three lines in the region 44-50 Teslas. There is a small peak at 28.5 Teslas resulting from the 200 K g-scattering phonon.

In order to determine the nature of the processes giving rise to these oscillations the series fundamentals were plotted as a function of applied uniaxial stress. This is shown in Fig. 1. The dependences of the series fundamentals upon stress were calculated from equation (16) and the known stress dependences of the donor binding energies, and are also plotted in Fig. 1. This figure demonstrates clearly the excellent fit which is given by the assumption that the oscillations arise from impurity capture at the various 1s donor states. The 23.5 T (X = 0) series rises rapidly with stress and seems to result from several unresolved series up to 1 kbar where one component splits off which is clearly attributable to a 685(f) - 1s(A_1) $[\propto^-_{11,21}]$ transition.

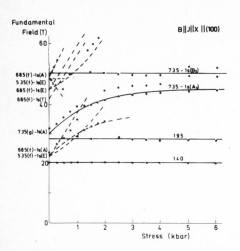

Fundamental Field (T)

B||J||x ||(100)

685(f)-1s(A)
535(f)-1s(E)
685(f)-1s(E)
685(f)-1s(T)

735(g)-1s(A)

685(f)-1s(A)
535(f)-1s(E)

735-1s(B₂)
735-1s(A₁)
195
140

Stress (kbar)

Fig. 1.
The variation of the experimentally determined fundamental fields for the magnetophonon series in silicon at 30K plotted as a function of applied uni-axial stress (j//B//X//100). The energy relaxation mechanisms thought to be responsible for the observed magneto-phonon series are marked with the pho-non energy in degrees Kelvin and the donor ground state into which capture occurs. The stress dependences of the g-scattering processes are plotted as solid lines. The theoretical dependen-ces of the f-scattering processes, which are supressed by stress, are plotted as dashed lines. In addition intravalley scattering from an optic phonon (753 K) is thought to give rise to the two sets of points 3T above the 735-1s(A₁) and 735-1s(B₂) points.

The weak series at 31.8 T (X = 0) rises in fundamental field and grows in amplitude with increasing stress. This process is very well des-cribed as a 735(g) to 1s(A₁) $[\propto^-_{11,21}]$ transition. The structure which appears at 40 T (X = 0) does not shift and is thought to result mainly from the second harmonic of the 20 Tesla series.

The application of only weak stresses is sufficient to remove the structure at 49 T (X = 0) providing confirmation that this series re-sults from the transverse valleys which relax by the same processes which give rise to the 23.5 T (X = 0) series. At higher stresses a further line appears at 50 Teslas which is well explained as a 735(g)--1s(B₂) transition which is enhanced, due to its g-like nature, by the stress.

It is thus concluded that at high stresses the energy relaxation is dominated by the emission of 140 K (g) phonons and of 735 K (g) phonons associated with electron capture at the 1s(A₁) and 1s(B₂) donor states. The assignment of the 735 K phonon as a Δ'_2 phonon has been made since this phonon is the only allowed g-scattering process at this energy (Streitwolf 1970). This value of energy compares well with the value measured by Dolling (1963) of 737 K. In addition there are 721 K Δ_5 g-phonons and a 753 K Γ'_{25} k=0 intravalley phonon both of which are group theoretically forbidden. One of these phonons is thought to give rise to two weak satellite lines which appear ~ 3 T above the two main peaks in the Fourier spectra which correspond to the 735 K phonon. These satellite lines are marked in Fig. 1 and correspond to a phonon energy of 755 K. This implies the presence of some scattering by the 753 K intravalley optical phonon, the energy of which has been

taken from the low temperature Raman measurements of Hart et al (1970).
Such a contribution would not be entirely unexpected, despite the
group theoretically forbidden nature of this process, in view of the
scattering present from the apparently forbidden 140 K g-type phonon.
Thus the selection rule forbidding intravalley optical phonon scattering, which has been derived in the absence of magnetic field, should
be applied with caution, at least in the interpretation of hot-electron data. Moreover, calculations by Laewatz (1969) and Tolpygo (1963)
suggest that intravalley optical phonon scattering may indeed be important if multipole terms are taken into account.

The general conclusions of the stress dependent magnetophonon experiments are given in Tables 4 and 5. Table 4 is a simplification of

Table 4. A simplified interpretation of the zero stress magnetophonon series in the light of the stress dependent results

Orienta-tion	Series fundamental fields (Teslas)	m^*	Resonance energy (K)	Calculated energy (K)	Capture process with phonon and final state
B ∥ (100)	23.5	m_1	164	157	685(f) — 1s(A)
				157	535(f) — 1s(E)
	31,8	m_1	220	207	735(g) — 1s(A)
	40	m_1	275	292	685(f) — 1s(T)
				307	685(f) — 1s(E)
	49	m_2	158	157	685(f) — 1s(A)
				157	535(f) — 1s(E)
	62	m_2	200	207	735(g) — 1s(A)
	93	m_2	299	292	685(f) — 1s(T)
				307	685(f) — 1s(E)
B ∥ (110)	30.5	m_1	166	157	685(f) — 1s(A)
				157	535(f) — 1s(E)
	64	m_1	206	207	735(g) — 1s(A)
B ∥ (111)	34.5		153	157	685(f) — 1s(A)
				157	535(f) — 1s(E)
	42		186	207	735(g) — 1s(A)
	61		265	292	685(f) — 1s(T)
	67		295	307	685(f) — 1s(E)

Table 2 in the light of the stress-dependence of the magnetophonon
series and gives the processes which probably give rise to the magnetophonon series. It is found that there is no process which clearly involves an f-phonon of energy 585 K. This phonon was only observed at
higher temperatures (Portal et al 1974, Eaves et al 1975) for series
resulting from the higher mass transverse valleys and gave a weaker
contribution than the 535 K f-phonon. There are however contributions

Table 5. The magnetophonon series present in the high stress limit for
$B \parallel J \parallel X \parallel (100)$

Series fundamental field (Teslas)	Resonance energy (K)	Calculated energy from eq.(20) (K)	Capture process with phonon and final state	Phonon energy, mode and symmetry from Dolling (1963)
inter Landau level series				
20.4	140			134 (TA, Δ_5)
28.5	195			221 (LA, Δ_1)
impurity capture series				
45.0	308	311	735(g) - 1s(A_1)	737 (LO, Δ'_2)
49.6	340	345	735(g) - 1s(B_2)	737 (LO, Δ'_2)
		331	755(Γ) - 1s(A_1)	755 (LO,TO, Γ'_{25})
52.5	359	365	755(Γ) - 1s(B_2)	755 (LO,TO, Γ'_{25})

from all of the processes given in Table 1 for the 535 K (f) and
735 K (g) phonons with the exception of 535 K - 1s(T), whose resonance
energy (143 K) is obscured by the 140 K (g) phonon, and of 735K(g)-1s(E)
and 1s(T) which are weak processes for X = 0. Capture to the high
stress analogue of the 1s(T) state is however observed for the 735K(g)
phonon at high stresses. This is shown in Table 5 which gives the magnetophonon series which are present in the high stress limit for
$B \parallel J \parallel X \parallel (100)$.

3. THE MAGNETO-IMPURITY EFFECT

Since the first investigation of the magnetophonon effect in the
warm electron regime (Stradling and Wood, 1968), hot electron experiments have shown a large number of complex features which result from
resonant energy relaxation. The initial observations were with n-GaAs
and subsequent experiments included some with n-InSb, n-InAs, n-CdTe,
N-Ge, n-Si and p-Te (see review by Harper et al 1973). The most widely
investigated of the additional structure arises from the emission of
high energy phonons accompanied by capture at the 1s ground state of
the impurity as discussed for n-Si in the previous section. Under warm
electron conditions magnetophonon structure has also been found to
arise from the emission of pairs of T.A. phonons, normally located
close to the X-point where there is a particularly high density of
phonon states. A further process has been identified recently where the
resonance, instead of involving a phonon, occurs between the bound
states of impurities and free carrier Landau states so that

$$\Delta E_I = N\hbar\omega_c \qquad\qquad (1)$$

This so-called "magneto-impurity effect" has been reviewed by Eaves and Portal (1978, 1979). The most extensive observations of the effect have been with germanium (Gantmakher and Zverov 1976, 1977) where the resonance energy has been shown to change with binding energy of the impurity present in the particular sample studied as expected from equation (1). When more than one type of impurity was present in a sample Nicholas and Stradling (1978) were able to detect "central cell structure" on the magneto-impurity resonance in n-InP with similar resolution to that attainable by far-infrared spectroscopy of the impurities concerned (an excellent example of the fact that zero is a perfectly respectable spectroscopic frequency if the d.c. measurements involve a resonance such as the magnetophonon or magnetoimpurity effect). The exact nature of all the mechanisms which can give rise to the magnetoimpurity effect is not completely understood nor why in one sample the dominant energy should be the energy difference between the 1s and 2p states whereas in another the impurity binding energy should be involved. Initially it was thought that the free carriers in the Landau states lost energy by resonant impact excitation of electrons from the 1s to 2p_ states or by impact ionisation of the impurity (i.e. that there was a resonant "cooling" of the free carriers). However subsequent measurements of the third harmonic component of the current when an a.c. voltage was applied to the sample suggested another mechanism. The third harmonic component is a direct measure of the degree of carrier heating and experiments with n-InP (Nicholas and Stradling 1978) indicated that the majority of the transitions take place in the opposite direction to that initially postulated in what can be described as a "resonant Auger emission". In this process bound carriers lose energy thereby producing resonant heating of the free carriers.

4. MAGNETOPHONON AND MAGNETO IMPURITY-RESONANCES IN n-TYPE CdTe

Up until now there have been only brief reports of the hot-electron magnetophonon-like structure in n-type CdTe (Stradling et al 1970, Harper et al 1973), which was interpreted before the shallow donor energies were accurately determined. This section summarises a detailed investigation of the hot-electron oscillatory structure in several samples of n-type CdTe where it is found that the different impurity concentrations give rise to different oscillatory structure.

Figure 2 shows the structure observed with the purest sample
C.N.R.S.1, at temperatures between 20K and 11K. In this sample the
peak mobility was 14.6 m^2/Vs. At 20 K the structure is dominated by a
series of oscillations (series A) which appear as sharp minima of
comparable amplitude in the magnetoresistance in both the longitudinal
and the transverse configurations. The oscillations are closely
periodic in 1/B and have a fundamental field of 7.7 T. As the temperature falls to 14 K the oscillations change in form and a series of
maxima appears at low magnetic fields with a fundamental field of
$\sim$ 1.3 T (series B). By 11 K it is clear that a third series with a
fundamental field of 16.3 T (series C) has taken over.

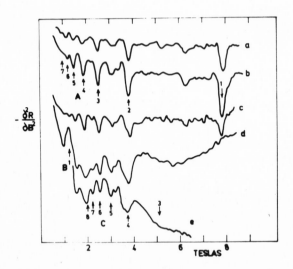

Fig. 2. Experimental recordings of derivative resistance against magnetic field for
a high purity sample of n–CdTe showing warm–electron magnetophonon structure. Curves a –e are taken in order of decreasing temperature in the range
20 to 11 K. Three distinct magnetophonon series labelled A–C appear in turn
as the temperature is lowered.

In the case of a less pure sample (μ_{peak} = 6 m^2/Vs) no oscillations were observable at 20 K. At lower temperatures there is a clear
series of maxima in the magnetoresistance (series D), with a periodicity
$(\Delta 1/B)^{-1}$ = 6.4 T. In addition there are two stronger maxima in resistance, or minima in conductivity, at 3.6 T and 9.5 T and two weaker
resitance, maxima at 7 T and 13.5 T.

Series A corresponds to an inter-Landau level series with a resonance energy of 71 cm^{-1}. This energy is a well known 2 T.A. phonon combination in CdTe, and the infra-red absorption experiments and shell model calculations of Batalla et al (1977) suggest that this is a pair of X-point phonons.

Series B is fitted by an impact ionisation/Auger process, which occurs between the 2a and 3p_ levels of the shallow donor. The peaks in resistance are described by the equation

$$N\hbar\omega_c = \Delta E_{2s-3p_-}$$

Series C, which also begins to appear at 14 K, becomes more clearly resolved at 11 K, where resistance minima appear to correspond to resonance. The resonance energy corresponds to 144 cm^{-1} and it is believed that the process involves electron capture at the 2s state of the impurity with L.O. phonon emission so that

$$N\hbar\omega_c = \hbar\omega_{LO} - E_I^{2s} \quad (B)$$

It may be noted that if the direction of the process giving rise to series B is "Auger-like" then both series B and series C, which appear at the same temperature, involve electron capture to the 2s state. This behaviour is exactly analogous to that observed in GaAs at 60 K where both the "Auger-like" and the L.O. phonon-impurity capture series appear at the same temperature and both series involve capture to the donor 1s state (Nicholas and Stradling 1976).

There are, in addition to the extrema described above, other features at 6.3 T and 3.2 T which are not explained by any of the above processes. There are many other possible mechanisms which may give rise to extrema. The fundamental fields for Auger excitation (or impact ionisation) from the conduction band to 2p_ and 2s states occur at ~ 6.6 T and ~ 3.6 T, and may be a possible origin of these extrema.

The resonant energy relaxation mechanisms for the lower purity material are quite different. Series D, which is a series of resistance maxima clearly dominant at low magnetic fields, is well described by the now well established energy relaxation process of L.O. phonon emission accompanied by capture at the donor 1s state, so that

$$N\hbar\omega_c = \hbar\omega_{LO} - E_I(B)$$

However there is no clear explanation for the additional peaks at 3.5T,

7 T and 9.5 T other than the two Auger processes mentioned above.

The overall picture of the resonant processes thought to be present, together with a comparison of the experimentally determined peak positions with those predicted by equations (3),(4),(5) and (6) are given in Table 6.

Table 6. A Comparison of the observed and predicted resonance positions in CdTe

Resonance index N	2 T.A. (X) series A		$3p_-$ - 2s series B		L.O. - 2s series C		L.O. - 1s series D	
0	–	–	–	–			14.5	13.5
1	7.76	7.76	1.46	1.31			4.54	4.95
2	3.88	3.84	0.72	0.66	7.86		2.59	2.56
3	2.59	2.55	0.50	0.45	5.14	5.1	1.85	1.89
4	1.94	1.93	0.37	0.34	3.85	3.75	1.44	1.48
5	1.55	1.55	0.30	0.28	3.08	3.04	1.17	1.18
6	1.29	1.28			2.58	2.60	0.99	1.00
7	1.11	1.10			2.20	2.34		
8	0.97	0.97			1.95	2.03		
9	0.86	0.86						
10	0.78	0.77						

5. EXTRINSIC OSCILLATORY PHOTOCONDUCTIVITY IN CdTe

In considering the very different behaviour of the two types of sample of the same material, CdTe, it is informative to consider other experimental evidence. Oscillatory photoconductivity, which occurs due to the very high probability of L.O. phonon emission by electrons photo-excited well into the conduction band, was first studied in CdTe by Mears et al (1968). It was these measurements that first demonstrated resonant capture at the ground state of the impurity. A recent investigation by Carter (D.Phil thesis 1977) has shown that for the less pure material the photoconductive response is a minimum when an electron, which is excited from a 1s donor state into the conduction band, can emit a series of L.O. phonons and be captured direction into a 1s or a n = 2 (2s or 2p) excited state. At higher electric fields the capture direct to a 2s or 2p state becomes the stronger process. In contrast, for the purer material only capture to the 1s state appears to be a strong process, although the strength and modulation of the oscillatory structure appears much weaker than in the less pure material. This is illustrated in Fig. 3 which shows oscillatory photoconductivity in both the pure (CNRS) and impure (GE) material. The differences are due to the condition necessary for the observation of strong oscillatory photoconductivity

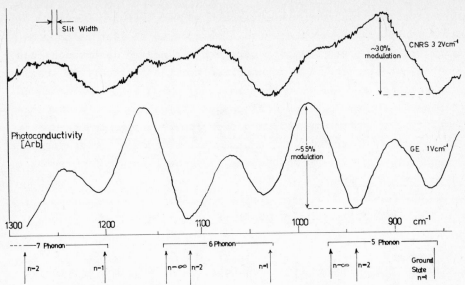

Fig. 3. Photoconductive response (unnormalised and in arbitrary units) against
frequency for two samples of n–CdTe of differring purity. Two series of
minima of different relative amplitude are observed; one is due to capture
at the impurity ground state (n=1) and the other due to capture at the
first excited state (n=2).

$$\tau_{L.O.} \ll \tau_{cp.} \ll \tau_{e.relax} = \tau_{a.c.}$$

where $\tau_{L.O.}$ is the L.O. phonon emission time, $\tau_{cap.}$ is the electron
capture time and τ_e is the energy relaxation time shich is determined
by acoustic phonon emission so that $\tau_e = \tau_{a.c.}$. In the G.E. material
$\tau_{a.c.} > \tau_{cap.}$, but due to the lower impurity concentration in the CNRS
material $\tau_{a.c.} < \tau_{cap.}$ so that the oscillatory structure is much
weaker. It should, however, be noted that the characteristic times used
in the relation above are not unique, due to the several different
processes which may occur. This is illustrated by the dominance of different photoconductivity minima for the two materials.

The behaviour observed in the magnetophonon and magneto-impurity
oscillations is the exact opposite to that found for the photoconductivity, with capture to the 2s state being dominant for the purer CNRS
1 and capture to the 1s state being much more pronounced for the G.E.
samples. This difference is not entirely surprising when it is realised
that the flow of electrons in energy space towards the resonance energy
is in the opposite direction for the two experiments. In the hot-electron magnetophonon effect the electrons are moving upwards in the conduction band under the influence of the electric field, whilst in the
photoconductivity experiments the electrons are falling down the band

following their initial photoexcitation. In both cases where the capture time is short the electrons will be captured as soon as they reach the relevant energy. As the carriers are heated up throught the band in the hot-electron magnetophonon effect the first strong process which is reached is capture to the 1s state, which will thus dominate for the less pure sample. Such a sharp cut-off will not influence the photoconductivity measurements. When the capture processes are weaker then the electron distribution function will extend to higher energies and other energy relaxation mechanisms, such as two phonon emission, will become significant.

6. EXTRINSIC OSCILLATORY PHOTOCONDUCTIVITY IN n-CdS

Observation of intrinsic oscillatory photoconductivity in CdS was first reported in 1964 by Park and Langer. Since then the intrinsic effect has received considerable theoretical and experimental attention (see Gastev et al 1973 for a list of references). The first observation of extrinsic oscillatory photoconductivity in n-Cd is reported by Carter (1977) and provides further evidence for the involvement of impurity states.

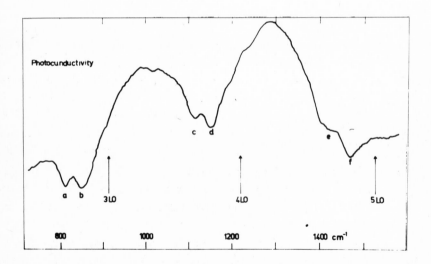

Fig. 4. Photoconductive response (unormalised and in arbitrary units) against frequency for n-CdS. The series of minima a,c, and e are due to capture at the n=2 impurity state with the emission of L.O. phonons; the series b,d and f are due to capture either at the n=3 state or in the tail states associated with the conduction band.

Figure 4 shows a typical unnormalised recording of the photoconductivity, made at 6 K with 2.5 Vcm^{-1}. The main features in the photo-
-response are labelled a-f. The b-d and the d-f spacings are very
similar, and closely equal to the LO phonon energy of 305 cm^{-1}. The
frequencies of the minima a-f are given in Table 7.

Table 7. Frequencies of the minima observed
in the extrinsic OPC of n-CdS

Line	Frequency	Separation	Frequency
a	807 ± 5 cm^{-1}	a-b	42 cm^{-1}
b	849 ± 5 cm^{-1}	c-d	46 cm^{-1}
c	1104 ± 5 cm^{-1}	e-f	64 cm^{-1}
d	1155 ± 5 cm^{-1}		
e	1400 ± 7 cm^{-1}	b-d	306 cm^{-1}
f	1464 ± 5 cm^{-1}	d-f	309 cm^{-1}

Henry and Nassau (1970),
by studying the "two-electron"
exciton recombination lines,
deduced that the binding energy
of a typical donor in CdS was
264 cm^{-1}, with an n = 2 energy
of 195 cm^{-1}. If the minima b,
d, f are to be explained in
terms of relaxation to the
band minimum according to the
relation, an impurity binding
energy of 240 cm^{-1} would be
required. The discrepancy of 24 cm^{-1} between this figure and the observed binding energy in CdS is very reasonable, as the tail of low
mobility states will extend considerably below the band edge in material of this purity. Alternatively impurity states with n $\geqslant$ 3 may
be involved. The energy of the n = 2 levels (195 cm^{-1}) is close to
that required (190 cm^{-1}) to explain the remaining minima a, c, e in
terms of capture at the 2s excited state i.e. $\hbar\omega = N\hbar\omega_{LO} + \Delta E_{1s-2s}$
(ΔE = 190 cm^{-1})

7. THE DONOR STATES IN GaP

The band structure of GaP has long been an area of experimental
and theoretical interest. The conduction band minima are believed to
be exceptionally non-parabolic because of the existence of a "Camel's
Back" form. Such a shape is expected to produce a strong non-parabolicity in the bands which is thought to extend over an energy range of
the same order as the binding energies of the p-like states of the
shallow donors normally fitted by effective mass theory. This results
in the breakdown of the relatively simple model of the donor states
associated with ellipsoidal energy surfaces appropriate to silicon and
germanium.

Furthermore, the theoretical analysis of donor levels in GaP is
complicated by the differences between the Ga and P site donors. Apart
from the site difference in symmetries, the Bloch part of the bound

electron wave function has nodes on the Ga sublattice and antinodes on the P sublattice (Morgan 1968). Therefore for excited states of a Ga site i.e. group IV donor effective mass theory would be expected to hold, since the total wave function vanishes at the impurity core. However for a P site i.e. group VI donor, the antinode in the Bloch function would accentuate any valley-orbit or central cell effects. This qualitative argument leads one to expect that effective mass theory could break down even for the excited states in the case of goup VI donors.

An extensive study has been made of the far-infrared absorption of the shallow donor states in GaP (Carter et al 1977). This study revealed many features which could not be explained in terms of the simple ellipsoidal model of the conduction band appropriate to the elemental semiconductors. In particular it was found that (i) there were small (~ 10 cm^{-1}) but significant differences in the binding energies of the p-states between different donors. (ii) The Zeeman splitting for the group VI substitutional donors S and Te was far smaller than predicted by the observed effective mass for the conduction band although the masses deduced from the Zeeman splittings for the group IV donor Si were close to the cyclotron resonance value. (iii) Some lines of the silicon substitutional donor which on an ellipsoidal model should be of even parity and hence should remain unsplit by a magnetic field, show a Zeeman splitting. (iv) There was a small systematic change in the effective masses deduced from the Zeeman splitting of transitions involving the silicon donor with the energy of the excited state concerned.

Although observations (iii) and (iv) might well arise from the highly non-parabolic form of the conduction band and GaP, (i) and (ii) clearly cannot be explained on any effective mass model of the impurity states concerned and remain a mystery. Although the model of states based on an ellipsoidal energy surface is not thought to give an accurate picture of any of the data, nevertheless it provides a convenient method of labelling the states concerned and all groups studying the donors in GaP have adopted this nomenclature.

Many unanswered questions therefore remain concerning the nature of the donor electronic states in GaP. In particular no exact interpretation of the observed spectra in terms of the "Camels's back" structure of the conduction band yet exists. Consequently the precise labelling of the states is yet unclear. Furthermore no consistency exists between the spacing of the lines observed with different donors in zero magnetic field and no explanation exists for the highly anomalous Zeeman splitting of the group VI substitutional donors.

A major new experimental approach has opened up as a result of the work of Scott (1979) who has identified "oscillatory photoconductivity" dips in the continuum photoresponse above 8000 cm^{-1} as arising from LO and LA phonon transitions back to the donor excited states. Figure 5 shows a recording taken from this work. Among the advantages implicit in this method of studying the excited states are that the photon energies involved are removed from the region of the strong combination phonon absorption bands and that, as phonons are involved, states of the same parity as the ground state become equally accessible spectroscopically.

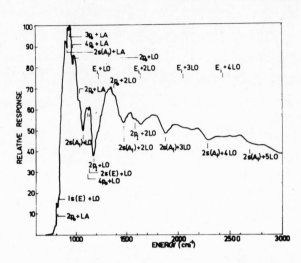

Fig. 5. Photoconductivity response against frequency for a sample of sulphur-doped GaP. Minima associated with capture at the 1s, 2s, 2p , 2p$_o$, 3p$_o$ and 4p$_o$ states accompanied by phonon emission are observed.

A further advance in the understanding of the donors in GaP is provided by the spectroscopic studies involving the application of calibrated uniaxial stress described in the thesis of Cooke (1979). By this means many of the weak absorption lines were shifted clear of obscuring phonon combination bands and an accurate value could be derived for the deformation potential Σ_u. The splittings observed for the sulphur donor are shown in Fig. 6. The reproducibility achieved for the splittings observed for the three lines 1s-2p$_+$, 1s-3p$_o$ and 1s-2p$_o$ is shown in Fig. 7 where the lines are labelled according to ellipsoidal band nomenclature. A deformation potential of $\Sigma_u = (6.3 \pm 0.7)$ eV can be deduced with the uncertainty arising from a lack of precise knowledge of the dimensions of the sample.

As a result of the recent measurements of Scott (1979) and of Cooke (1979) some consistency of the results for different impurities is now emerging although differences of the order of 5-10 cm^{-1} which are outside experimental error still remain from impurity to impurity.

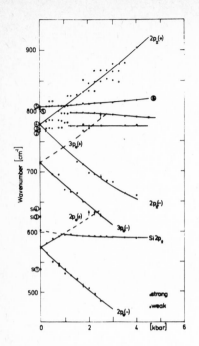
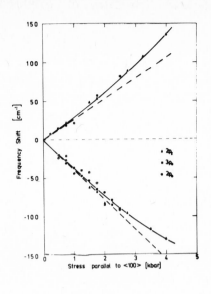

Fig. 6. Absorption peaks as a function of stress for a sample of sulphur doped GaP observed in an infrared transmission experiment.

Fig. 7. The change in frequency of absorption peaks in sulphur-doped GaP against uniaxial stress for the three absorption peaks corresponding to transitions from the ground state into the $2p_+$, $3p_0$ and $2p_0$ states.

The separation of individual lines from the strongest line which is taken to be the 1s-2p$_+$ state on the ellipsoidal interpretation is given in Table 8. The labelling of certain of states is different between Scott and the Oxford group and the two alternative interpretations on the ellipsoidal model are given in the table.

Table 8. A comparison of the separation of the donor lines in GaP from the strongest $1s$-$2p_{+1}$ line found for different donors by various experimental groups.

Identification		Si		S			Te		Se	O
Carter et al.	Scott	Theory (Carter et al.)	(Carter et al.)	(Carter et al.+ Cooke)	(Scott) LO	(Scott) LA	(Carter et al.+ Cooke)	(Scott)	(Scott)	Dean and Henry
$2p_{+1}$-$2p_0$		$191\ \mathrm{cm}^{-1}$	In restrahl	201	204	214	216	217	200	189
	$2p_{+1}$-$2s(A)$				109	114		119	106	
$2p_{+1}$-$3p_0$	$2p_{+1}$-$3p_0$			93		89	92	89		77
$2p_{+1}$-$4p_0$	$2p_{+1}$-$4p_0$	64	64	62	60	58	65	70	60	49
	$2S(E)$-$2p_{+1}$				49				45	
$2p_{+1}$-$4p_0$	$3s(E)$-$2p_{+1}$	9.4		12, 7	8					
$4f_0$-$2p_{+1}$	$5s$-$2p_{+1}$	22.7	22.3	29	26					28
$3p_{+1}$-$2p_{+1}$		36	36							

ACKNOWLEDGEMENTS - The experimental work described in this paper was mainly performed by Drs. A.C.Carter, R.A.Cooke and R.J.Nicholas in preparation for their D.Phil. theses at Oxford University.

REFERENCES

1. Aggarwal R.L. and Ramdas A.K. (1965) Phys.Rev. 140, A1246
2. Anastassakis E., Pinczuk E. and Burstein E. (1970) Solid State Commun. 8, 133
3. Carter A.C., Dean P.J., Skolnick M.S. and Stradling R.A. (1977) J.Phys. C10, 5111
4. Carter A.C. (1977) D.Phil.Thesis, Oxford University
5. Cooke R.A., Nicholas R.J., Stradling R.A., Portal J.C. and Askenazy S. (1978) Solid State Commun. 26, 11
6. Cooke R.A. (1979) D.Phil. Thesis, Oxford University
7. Dolling G. (1963) Inelastic Scattering of Neutrons in Solids and Liquids 2 37 I.A.E.A.
8. Eaves L., Hoult R.A., Stradling R.A., Askenazy S., Barbaste R., Carrere G., Leotin J., Portal J.C. and Ulmet J.P. (1977) J.Phys.C: Solid St. 10, 2831
9. Eaves L., Hoult R.A., Stradling R.A., Tidey R.J., Portal J.C. and Askenazy S. (1975) J.Phys.C: Solid St. 8, 1034
10. Eaves L., Portal J.C., Askenazy S., Stradling R.A. and Hansen K. (1974) Solid State Commun. 15, 1281
11. Gantmakher V.F. and Zverev V.N. (1976) Sov.Phys. JETP 43, 985
12. Gastev S.V., Lider K.F. and Novikov B.V. (1973) Sov.Phys.Semicond. 7, 613
13. Gureivich V.L. and Firsov Y.A. (1961) Sov.Phys.JETP 13, 137
14. Hardy J., Smith S.D. and Taylor W. (1962) Proc.Int.Conf. on Phys. of Semiconductors (Exeter) p521.
15. Harper P.G., Hodby J.W. and Stradling R.A. (1973) Rep.Progr.Phys. 37, 1
16. Hart J.R., Aggarwal R.L. and Lax B. (1970) Phys.Rev. B1, 638
17. Hensel J.C., Hasegawa H. and Nakayama M. (1965) Phys.Rev. 138,A225
18. Henry C.H. and Nassau K. (1970) Phys.Rev. B2, 997
19. Kohn W. and Luttinger J.M. (1955) Phys.Rev. 98, 915
20. Lax M. and Birman J.L. (1972) Phys.Status Solidi B49, K153
21. Mears A.L., Spray A.R.L., Stradling R.A. (1968) J.Phys. C1, 1412
22. Nicholas R.J., Carter A.C., Stradling R.A., Portal J.C., Houlbert C. and Askenazy S. (1979), to be published
23. Nicholas R.J. and Stradling R.A. (1976) J.Phys.C: Solid St. 9, 1253
24. Nicholas R.J. and Stradling R.A. (1978)
25. Ousset J.C., Leotin J., Askenazy S., Skolnick M.S. and Stradling R.A. (1976) J.Phys.C: Solid St. 9, 2803
26. Palevsky H., Hughes D.J., Kley W. and Tunkelo E. (1959) Phys.Rev. Lett. 2, 258
27. Portal J.C. (1975) D.Phil. Thesis University of Toulouse
28. Portal J.C., Eaves L., Askenazy S. and Stradling R.A. (1974) Solid State Commun. 14, 1241
29. Scott W. (1979) J.Appl.Phys. 50, 472
30. Stocker H.J. and Kaplari M. (1966) Phys.Rev. 150, 613
31. Stradling R.A. and Wood R.A. (1970) J.Phys. C3, L94 and 2425 also Proc. Int. Conf. on Physics of Semiconductors (Boston) p.369
32. Stradling R.A. and Zhukov V.V. (1966) Proc.Phys.Soc. 87, 263
33. Streitwolf H.N. (1970) Phys.Status Solidi 37, K47
34. Tekippe V.J., Chandrasekhar H.R., Fisher P. and Ramdas A.K. (1972) Phys.Rev. B6, 2348
35. Wallace P.R. and Joos B. (1978) J.Phys.C: Solid St. 11, 303
36. Wilson D.K. and Feher G. (1961) Phys. Rev. 124, 1068

OPTICALLY DETECTED MAGNETIC RESONANCE
STUDIES OF SEMICONDUCTORS

B.C. Cavenett
Department of Physics,
University of Hull,
Hull, U.K.

This paper briefly reviews the application of optically detected magnetic resonance (ODMR) to the study of electron-hole recombination processes in semiconductors. The high sensitivity of the technique and the fact that the resonances can be directly linked to particular emission processes allows detailed characterization of exciton, donor--acceptor and deep trap recombination. Observations of bound and free triplet excitons in GaSe and GaS are reported and details of deep acceptor centres associated with vacancies in II-VI compounds are given. A recent investigation of oxygen in GaP confirms the existence of the two electron trap and studies of amorphous silicon show that the ODMR technique can be used to study deep centres in both crystalline and amorphous semiconductors.

INTRODUCTION

The study of defects in semiconductors by magnetic resonance has continued for many years, but more recently the interest in deep traps which control optical and electrical properties of devices has produced a resurgence of interest in the technique. For example, Kaufmann, Schneider and Rauber (1) have identified a phosphorus anti-site centre, PP_4, in GaP and the gallium vacancy, V_{Ga}, has been observed by Kennedy and Wilsey (2). In the latter case, the resonances were only observed when the sample was illuminated with light, illustrating the importance of photo-excited magnetic resonance experiments. Krebs and Stauss (3) have observed the various charge states of chromium in GaAs, namely $Cr^+(3d^5)$, $Cr^{2+}(3d^4)$ and $Cr^{3+}(3d^3)$ using similar techniques.

Magnetic resonance is an important technique because the observation of microwave transitions between the Zeeman levels of a defect can provide very detailed information about the charge state, the site symmetry, covalency, and pairing of defects. Interactions between the electron spin and the nuclear spins of the defect and the neighbouring atoms can confirm the identity of the impurity and the precise lattice location, that is substitutional, off-centre or interstitial sites. For details see Abragam and Bleaney (4) who review the application of the technique to transition metal ions in solids.

It is perhaps, at first, surprising that the use of magnetic resonance in semiconductors is not more widespread, particularly when it is clear that the technique complements other methods such as luminescence, Zeeman spectroscopy, photoconductivity and photocapacitance for investigating the role played by defects in electron-hole recombination processes. There are two main reasons for this. The first is sensitivity since conventional magnetic resonance is limited in practice to $\sim 10^{15}$ cm^{-3} spins since only small samples can be used and line widths are often several hundred gauss. In high quality semiconductors the defect levels will be of this order or less and so resonances are difficult to observe without deliberate doping or by radiation damage in the case of vacancy centres. Also it should be noted that in pure semiconductors many important "defects" are created by excitation of the crystal by light. These excited states of the crystal such as free and bound excitons are usually short lived and insufficient in number to be observed by conventional magnetic resonance. The second reason depends on the fact that magnetic resonance will be detected from *all* of the paramagnetic centres in a sample irrespective of whether they are important in electron-hole recombination. Therefore, it is not often clear whether the centres observed in magnetic resonance are related, say, to the levels observed in transient photo-capacitance. Photo-induced resonance experiments can often clarify the situation but, in general, this difficulty of assignment and the lack of sensitivity limits considerably the application of magnetic resonance to the study of pure semiconductors.

Optically detected magnetic resonance (ODMR) was developed by Geschwind et al (5) in order to investigate excited states of magnetic ions in insulators and, although there was initially little interest in the application of the technique to semiconductors, recent results have shown that the high sensitivity of the technique can allow the resonances to be linked directly to particular emission processes. In this paper the ODMR technique is briefly discussed and examples are given from the studies of excitons, donor-acceptor pairs and deep traps studied in this laboratory. Further details can be found in several reviews by the author (6),(7),(8) .

OPTICALLY DETECTED MAGNETIC RESONANCE

The optical detection of magnetic resonance depends on the fact
that microwave induced transitions between the Zeeman levels of the
excited or emitting state of a system can change the total intensity
of the emission or the intensities of the circularly polarized emission
components. As in conventional magnetic resonance the sample is placed
in a microwave cavity but instead of the power absorbed at resonance
being detected by a microwave bridge, luminescence is excited by a
laser and at resonance the change in emission intensity is detected.
Whether the ODMR signals are obtained as changes of the total intensity
or changes in the circularly polarized emission components depends on
whether the system is thermalized or not, that is, whether or not the
spin lattice relaxation time, T_1, is much shorter than the optical
decay time, γ. These two cases are illustrated in Fig. (1) for donor-
-acceptor (D-A) recombination. In the case (a) which is applicable to

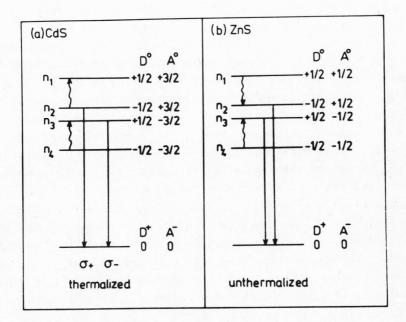

Fig. 1. Optical resonances in (a) CdS where the system is thermalized and the donor
resonance is detected as changes in the polarized emission components, and
(b) ZnS where donor and acceptor resonances are observed as changes in the
total emission intensity for an unthermalized system.

the green edge emission in CdS, the excited state is that of a D-A
pair with the allowed electron-hole recombination transitions shown
for B// c-axis. If we consider the observation of the donor resonance,
then we note that because the system is thermalized $n_2 > n_1$ and $n_4 > n_3$.
At resonance, the intensities of the circularly polarized emission
components change with σ^+ decreasing and $\bar{\sigma}$ increasing. Thus, for
this system the experiment must be carried out by monitoring either
$I_{\sigma+}$ or $I_{\sigma-}$ in a direction parallel to the magnetic field. In the second
example, Fig. (1b) the energy scheme is shown for unthermalized D-A pairs
in ZnS where the deep acceptor is described by $S = \frac{1}{2}$. The highly allowed
recombination transitions result in the populations of n_2 and n_3
dropping below those of n_1 and n_4. If we again consider the donor resonance, transition take place from $n_1 \rightarrow n_2$ and $n_4 \rightarrow n_3$, increasing
both emission components. Thus a microwave induced change in the total
emission is observed.

Two types of problems are of interest in studies of semiconductors.
On the one hand, it is important to investigate the magneto-optical
properties of free and bound carriers since the electronic properties
of these systems are determined by the band structure of the material.
Then magnetic resonance investigations of electron, hole and exciton
g-factors provide tests for the approximations used in band theory
calculations. As we have already discussed in the introduction it is
also important to investigate the electronic structure of defects and
impurities in semiconductors. Thus magnetic resonance measurements
enable detailed investigations of vacancy, interstitial and impurity
centres. The principal parameter in the resonance is the g-factor since
the simplest expression that can be written is $h\nu = g \mu_B B$. We can
summarize the two types of problems by noting that in the first examples given above the g-factors are determined by the band structure.
Thus, in the case of donors, shallow acceptors and bound excitons the
nature of the centre binding the charged carriers cannot be determined
from the resonance information. However, in the second set of examples,
the g-factor and other parameters do reflect the local charge state,
symmetry and nuclear interactions and so identification of the centre
is generally possible.

An ODMR spectrometer using a superconducting magnet is shown in
Fig. 2. The samples are placed in a microwave cavity (9 GHz or 16.5 GHz)
at the centre of a 2.5T superconducting magnet. Luminescence from the
sample is excited by either an argon or krypton ion laser and the
emission is monitored by a photomultiplier in a direction parallel to
the magnetic fields so that microwave induced changes in either

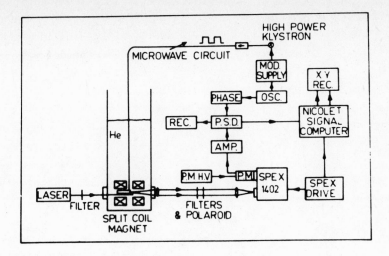

Fig. 2. Optical magnetic resonance spectrometer

the total intensity or one of the circularly polarized components can
be detected. Up to 16 watts of microwaves are provided by a klystron
and a travelling wave tube amplifier. The microwaves are switched at
audio frequencies and the resonances are detected using a conventional
lock-in operating at the microwave chopping frequency. A signal
averager can be used to enhance the signal-to-noise. The sign of the
emission change is measured by comparing the wave form of the chopped
microwaves with the wave form of the signal taken directly from the
photomultiplier and accumulated on the signal averager. Since the
photomultiplier output is negative, wave forms in phase correspond to
a decrease in the emission intensity and out of phase wave forms correspond to an increase in emission. In order to determine which emission lines or bands change in intensity at resonance a spectral dependence measurement is made by setting the magnetic field to the resonance and using a spectrometer to analyse the luminescence. Thus a recording to ΔI versus wavelength is obtained. This is the arrangement
shown in Fig. 2.

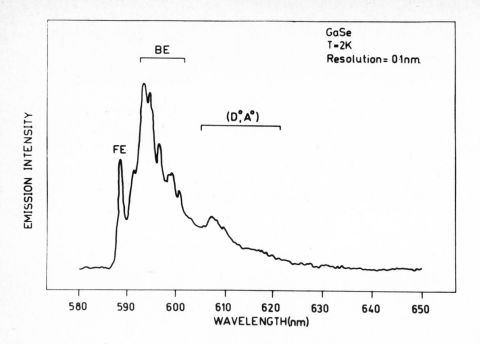

Fig. 3. Emission spectrum for the layered semiconductor GaSe showing the free exciton (FE), bound exciton (BE) and donor-acceptor luminescence.

EXCITON RECOMBINATION IN LAYERED SEMICONDUCTORS

GaSe

Recently, the first observation of a bound triplet exciton in the semiconductor GaSe was reported by Dawson et al (9) and Morigaki et al (10). GaSe is a layered semiconductor with axial symmetry and the ODMR experiments were carried out with B// c-axis and observation of the luminescence parallel to B. The emission is shown in Fig. (3).

Two resonances were observed from the bound exciton emission region, a low field increase in the σ^+ emission and a high field increase in the σ^- emission. These resonances which are shown in Fig. (4) are characteristic of an unthermalized triplet exciton with an energy level scheme as shown in Fig. (5). The populations of the $|1>$ and $|-1>$ states drop below the population of the $|0>$ state because of the larger decay rate out of the $|\pm1>$ states. The Hamiltonian describing the system can be written

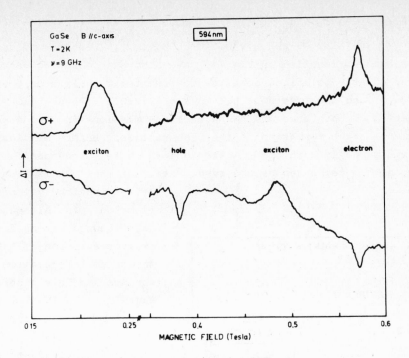

Fig. 4. Optical resonance at 594 nm for GaSe with B// c-axis showing the triplet exciton, hole and electron resonances.

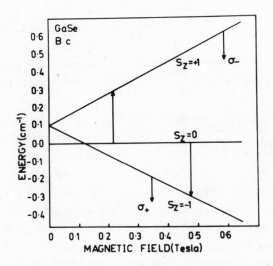

Fig. 5. Energy level scheme for the triplet exctions in GaSe. The microwave transitions shown correspond to an unthermalized system.

$$\mathcal{H} = g_{ex}// \mu_B B_z S_z + g_{ex} \mu_B (B_x S_x + B_y S_y) + D[S_z - (1/3)S(S+1)]$$

where $g_{ex}// = 1.85 \pm 0.03$ and $D = 0.110 \pm 0.004$ cm^{-1}. Electron and free hole resonances, labelled in Fig. 4, have also been observed as changes in the circularly polarized emission components. The electron g-value is 1.13 ± 0.01 and the free hole has $g_h// = 1.72 \pm 0.02$. These resonances are observed on both free and bound exciton emissions and are due to spin memory in the formation of the free excitons, before binding as shown in Fig. 6. At both the electron and the hole resonances I_{σ^+} increases and I_{σ^-} decreases as observed. Note that the hole is labelled

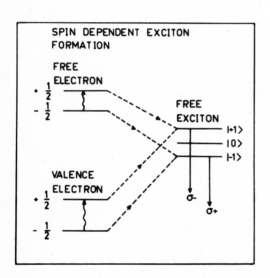

by the unpaired valence electron spin. Although GaSe shows a strong free exciton emission at 588 nm, so far no free exciton resonance has been observed. Magnetic circular dichroism measurements have also been carried out as a function of magnetic field strength. A large signal is observed at 1.5 kgauss corresponding to the crossing of the $|0>$ and $|-1>$ bound exciton levels. This value is in good agreement with the observed g-value and zero-field splitting obtained from the ODMR experiment.

Fig. 6. Model showing that the electron and hole resonances are observed because of spin dependent exciton formation from thermalized electrons and holes.

It was surprising to discover that there are two exciton spectra in GaSe corresponding to different emission spectra from the samples. The type I spectrum has been described above and is characterized by a sharp free exciton emission at 488 nm. In the type II material this emission is very weak ($\sim 1/200$) and the bound exciton lines are less resolved. The ODMR signals are much stronger in type II material and the triplet spectra are described by the same g-value as type I but with $D = +0.288 \pm 0.004$ cm^{-1}. The level crossing occurs at a correspondingly larger magnetic field. Schmid et al (11) and Mercier and Voitchovsky (12) first showed that the luminescence spectra were made up of two spectra corresponding to different extreme doping levels. However, the ODMR data does not clearly correlate with doping concentration since only the two resonance spectra have been observed and usually only one resonance is observed in each sample. Cavenett et al (13) have suggested that the two resonances are associated with the different polytypes, γ and ε.

GaS

GaS is also a layered compound but has an indirect gap well below the direct gap and exhibits only one polytype, β. The emission spectrum is shown in Fig.(7) with the high energy edge shown in the

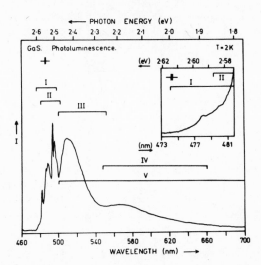

Fig. 7. Emission spectrum for GaS with the high energy edge shown in the inset. The labels I to V indicate the emission regions associated with the exciton resonances shown in Fig. 8.

inset. Five exciton resonances have been observed in this material as shown in Fig.(8). The resonances are labelled I to V and the relevant spectral regions are shown and labelled in Fig.(7). All of the excitons have the same g-value, $g_{ex//} = 2.006 \pm 0.002$, but the zero field splittings are $D^I = 0.013$ cm^{-1}, $D^{II} = 0.024$ cm^{-1}, $D^{III} = 0.025$ cm^{-1}, $D^{IV} = 0.075$ cm^{-1} and $D^V = 0.010$ cm^{-1}. The resonances from the high energy wing are remarkably narrow and Dawson et.al. (14) have attributed these resonances to free indirect excitons. A free exciton moving through the lattice is subject only to the average crystal field and

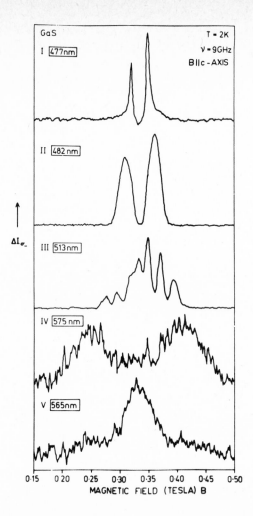

Fig. 8.   Optically detected resonances of excitons in GaS. Type I resonance
has been attributed to free indirect excitons and types II-V are
indirect bound at different sites in the lattice.

hyperfine interactions; in both cases the fluctuations average to zero
and the resonances are described as motionally narrowed. The zero field
splitting of the triplet state will correspond to that due to the un-
distorted lattice and the widths of the resonances will be determined
either by the spin lattice relaxation time or the optical decay time.
The four other resonances have been attributed to excitons bound at
different sites in the lattice; for example type III shows strong interaction with one Ga nucleus.

DONOR-ACCEPTOR RECOMBINATION

CdS

Recombination takes place between an electron described by $m_S = \pm \frac{1}{2}$ on a donor and a hole, $m_J = \pm 3/2$, on a shallow acceptor. Fig. 1a shows that for a thermalized pair the donor resonance changes the intensities of the circularly polarized emissions. The ODMR signals are shown in Fig. (9). The CdS emission and the spectral dependence of the resonance are shown in Fig. (10) where it can be seen that only the D-A pairs (and LO replicas) contribute to the resonance; the free electron-to-acceptor transitions (and LO replicas) do not contribute. The mea-

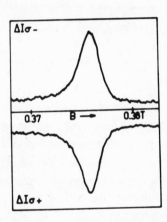

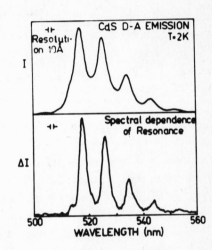

Fig. 9. Polarization dependent donor resonances in CdS. For energy level
scheme see Fig. 1a.

Fig. 10. Green donor-acceptor luminescence in CdS (top) and spectral
dependence of the donor resonance showing that the free-to-
-bound transitions do not contribute to the resonance.

sured $g_e//$ is 1.789 ± 0.002 and at resonance the polarized emission components
change by ± 0.4 %. There was no sign of microwave saturation of the
signal. The calculated changes assuming the model in Fig. 1a and
assuming microwave saturation of the resonance show that maximum changes
of ± 10 % can be expected. Further details of this work can be found in
Brunwin et al (15) and Dunstan et al (16).

ZnS and ZnO

Crystals of ZnS excited with UV show many broad emission bands in
the blue, green and red regions. Investigations of these recombination
processes has been carried out by James et al (17),(18) and Nicholls

et al (19). Both donor and deep acceptor resonances were observed by ODMR and the results for ZnS are shown in Fig. (11). The high field resonance at g = 1.886 is that of the donor and the low field, broad re-

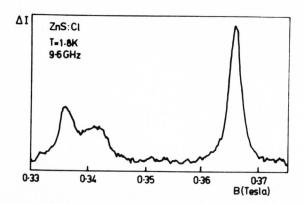

Fig. 11. Donor and acceptor resonances in ZnS. The acceptor resonances is that of $(V_{Zn}-Cl)$ centres.

sonance was identified as an A-centre acceptor. The A centre is a vacancy-donor complex such as $(V_{Zn}-Cl)$ and is formed when an electron is removed from the self-activated centre by UV. Spectral dependence measurements showed that both donor and acceptor resonances came from the 435 nm emission band so establishing the D-A nature of this emission. This change in the emission was measured to be 0.3 % and the results can be explained in terms of an unthermalized model as shown in Fig. (1b). Similar results have been observed in ZnO:Li by Block et al (20) and Cox et al (21).

D-A pairs are characterized by recombination rates which depend on the separation. For close pairs the recombination rate is higher than for distant pairs. Thus, in time resolved experiments the close pair emission will be sampled preferentially for short delays after the excitation light pulse. A time resolved ODMR system has been developed by Dawson et al (22) and used to investigate ZnS (23) and ZnO (24). The experiments were carried out by using a pulsed nitrogen laser at 337 nm for an excitation source. The laser was pulsed at frequency f and microwaves from a magnetron or a klystron and TWT were pulsed at f/2. The luminescence was sampled with a dual channel box-car unit such that the intensity of emission with microwaves on was recorded in one channel and the intensity without microwaves was recorded in the other channel. The ODMR signals were obtained by digitally subtracting the two channels and recording the output on a signal averager. The sample gate width was variable from 35 nsec to 3 msec and the gate delay ranged from 0 to 6 msec. Time resolved emission measurements could also be carried out by using the system as a single channel box car. The time resolved emission results for ZnS are shown in Fig. (12a) and the corresponding time resolved ODMR signals are given in Fig. (12b).

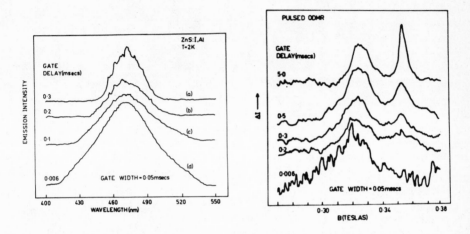

Fig. 12. a) Time resolved emission spectra for the self activated emission in ZnS.
b) Time resolved optical resonance spectra for donor-acceptor recombination in ZnS.

For long delays the time resolved ODMR is similar to the CW ODMR shown in Fig. (11). As the delay becomes short, the emission broadens and the donor and acceptor resonances merge together eventually giving one broad resonance at $g = 2.09 \pm 0.03$. This resonance which is an increase in the emission has been interpreted as that of exciton recombination at close D-A pairs. The total spin is $J = 1$ so that in a magnetic field the level scheme is as shown in Fig. (13). For close pairs with similar separations the zero field splittings and the g-values depend on the orientations of the pair with respect to the magnetic field direction, thus a broad resonance would be expected. Dawson and Cavenett (23) suggest that the so called Cu-blue emission is exciton recombination on nearest neighbour D-A pairs analogous to GaP:Cd,O (25).

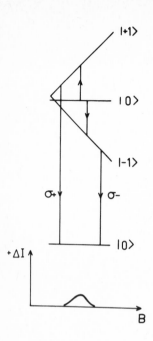

Fig. 13. Triplet model for close donor-acceptor
pair   resonances in ZnS corresponding to the signal in Fig. 12b
taken with a delay   time of G μsec.

ZnTe

The values of the elec-
tron g-value in ZnTe has been
uncertain because of very
different reported values.
Hollis et al (26) measured
g_e = 0.57 using spin-flip
Raman measurements but in
later magneto-optical measurements the value was revised
to 1.74 (27). Magneto-ref-
lectance studies by Venghaus
et al (28) on the free exciton in ZnTe gave a value of
g_e = -0.57 and Dean et al
(29) obtained a value of
-0.38 ± 0.05 by investigating
the Zeeman splitting of an
(A^o,X) emission line. Killoran
et al (30) have investigated
ZnTe crystal which showed
both (A^o,X) and (D^o,A^o) recombination in the 520-550 nm
region. The electron resonance at g_e = 0.401 ± 0.004 was observed by
ODMR as a change in the polarized emission components as shown in
Fig. (14). Spectral dependence measurements have not determined which of
the two recombination processes is responsible for the resonance but
(D^o,A^o) is most likely. This process is shown in Fig. (15) for the $S = \frac{1}{2}$
donor and J = 3/2 acceptor.

ZnSe

The electron g-value has been measured using ODMR by monitoring
the blue edge emission in ZnSe. The donor resonance gives
g_e = 1.115 ± 0.010 (31) but no shallow acceptor resonances were observed.
The energy level diagram shown for ZnTe (Fig. 15) also describes the
spin dependent recombination in ZnSe.

ZnSe doped with Cl, I or Cu shows many broad emission bands which
have been investigated by many techniques. Many of these bands are
thought to be donor-acceptor in nature and as in the case of ZnS both
donor and acceptor resonances have been observed from several emission
bands confirming the pair nature of these emissions. Dunstan et al (32)

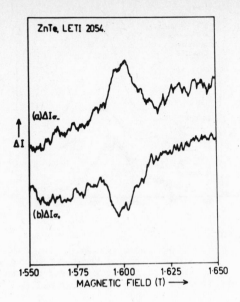

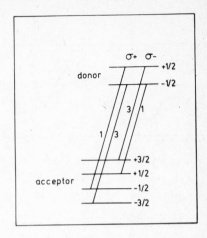

Fig. 14. Polarization dependent donor optical resonance in ZnTe.

Fig. 15. Recombination model for donor-acceptor recombination in ZnTe with the electron spin S=1/2 and shallow acceptor spin, J=3/2.

showed that the 632 nm emission in ZnSe:I was due to donor $-V_{Zn}^-$ recombination. The resonances are shown in Fig. (16) and the spectral dependences of the donor and the V_{Zn}^- resonances are shown in Fig. (17). Although all of the emission is associated with the donor resonance, only the 632 nm band is associated with the acceptor. Angular dependence studies confirmed that the acceptor was indeed the isolated zinc vacancy and the results were fitted using the g-values of Watkins (33). In ZnSe:Cl single crystal Nicholls et al (34) reported the observation of a donor-vacancy pair acceptor analogous to the A-centre in ZnS. The emission associated with this centre was determined to be at 620 nm. Assuming that the transitions involving these vacancy centres have a zero-phonon line at the high energy edge, the two levels can be placed at ~ 0.6 eV from the valence band. This value compares favourably with the 0.59 eV level observed in undoped ZnSe by Grimmeiss et al (35) by photocurrent and photocapacitance measurements. A full description of the ODMR investigations can be found in Dunstan et al (36).

More recently Davies and Nicholls (37) and Nicholls and Davies (38) have reported donor-to-P_{Se} acceptor recombination from the 1.91eV emission confirming that phosphorus on a Se site acts as a deep acceptor at about 0.6-0.7 eV from the valence band.

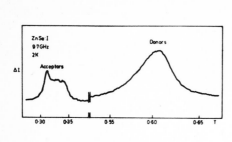

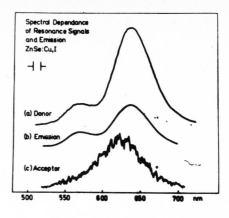

Fig. 16. Donor and acceptor resonances for ZnSe:I where the acceptors are zinc vacancies, V_{Zn}^-.

Fig. 17. Emission spectrum (b) for ZnSe; Cu, I showing the same resonances as in Fig.16. The donor resonance (a) is associated with all of the emission, but the acceptor resonance (c) comes only from the band peaking at 632 nm.

THE DEEP TRAP O⁻ IN GaP

Investigations by Dean et al (39) have shown that oxygen is a deep donor on a P site and the neutral donor binds an electron at 0.895 eV at 1.6 K. Oxygen donor to acceptor recombination results in near infra-red pair spectra ($\sim$ 1.2-1.5 eV). Dean and Henry (40) have shown that, in addition to the donor-acceptor transitions, an infra-red emission with a zero-phonon line at 0.84 eV and many phonon replicas is also due to oxygen. This luminescence was thought to result from a radiative transition of an electron from the 1s(E) state of the isolated oxygen donor to the 1s(A) ground state. The two electron oxygen centre, O^-, was proposed by Kukimoto et.al (41),(42) in order to interpret photo-capacitance measurements. They suggested that the donor, O^0, could capture a second electron with a subsequent lattice distortion. However, Grimmeiss et al (43) and Morgan (44) interpreted their two-electron photocapacitance data without involving a large Franck-Condon shift. Recently Morgan (45) reinterpreted the 0.84 eV emission data of Dean and Henry (40) suggesting that this emission is due to a radiative transition of the two electron oxygen state.

The 0.84 eV emission from an oxygen doped GaP sample is shown in Fig. (18). A typical ODMR spectrum from this O^- band is shown in the upper part of Fig. (19). The magnetic field is along the [110] direction

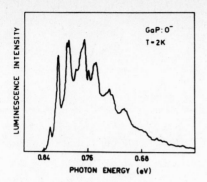

Fig. 18. Infra-red emission spectrum from
GaP:O⁻ with zero phonon line at
0.84 eV.

and the microwave frequency is
16.5 GHz. Angular dependence studies
showed that these resonances can
be accounted for by a spin-triplet
in the emitting state of the O⁻
centre which has axial symmetry
around the [110] direction and a
nuclear hyperfine interaction
with one Ga nucleus. The spin
Hamiltonian is

$$\mathcal{H} = g\mu_B B_z S_z + g\mu_B (B_x S_x + B_y S_y) +$$

$$+ D\left[S_z - \frac{1}{3}S(S+1)\right] + A_1 \underline{I} \cdot \underline{S} + A_2 \underline{I} \cdot \underline{S}$$

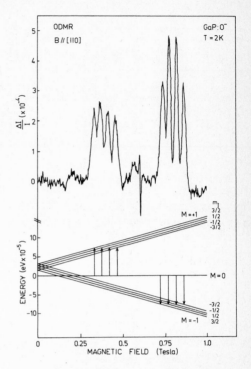

Fig.19. Optical resonance spectra from
the two electron oxygen centre
in GaP for B// [110] and at 16.5
GHz (upper). The lower part of
the figure shows the triplet
energy level scheme with an interaction with one Ga atom due
to the [110] distortion.

where S=1 and D is the zero field
splitting constant. For B// [110] g=2.011±0.005, D=2.32±0.01x10⁻⁵ eV,
$A_1(^{69}Ga)=4.40±0.02x10^{-6}$ eV, $A_2(^{71}Ga)=5.88±0.02x10^{-6}$ eV. The resonances
are increases in the luminescence and so the system is an unthermalized
triplet.   The energy level diagram is shown in Fig. (19) and the
level crossing measurements are shown in Fig. (20). The observed interaction between the O⁻ state and a single Ga nucleus with I=3/2 can be
explained if we suppose that when the O° captures a second electron
into the O⁻ excited state the lattice relaxes to a new equilibrium
state characterized by a [110] distortion. Such a spontaneous distortion can be due to the Jahn-Teller effect which lifts the orbital de-

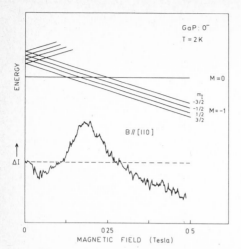

Fig. 20. Level crossing measurement for GaP:O⁻ showing the increase of ΔI as the M=0 level crosses the M=-1 level.

generacy of the triplet system. Thus, the ODMR measurements show that the 0.84 eV emission is due to an internal transition of the O⁻ centre, namely from a triplet excited state to a singlet ground state. Further details of this investigation can be found in Gal et al (46).

AMORPHOUS-Si

Since the nature of the recombination centres and mechanisms were not clearly understood, an investigation of a-Si was undertaken by Morigaki et al (47) using ODMR. As shown in Fig. 21 three resonances were observed from the total emission, two narrow lines with g = 2.006^{+}0.001 (D$_2$) and g = 2.018^{+}0.002 (D$_1$) which were decreases of the emission at resonance

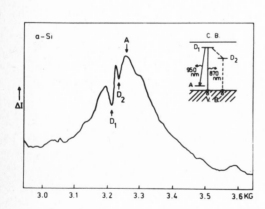

Fig. 21. Optical resonance spectra for a-Si showing the acceptor, A, and electron resonances D$_1$ and D$_2$. The inset shows the radiative (solid lines) and non-radiative recombination processes.

and a broad resonance at g = 1.999^{+}0.010 which was an increase in light. Spectral dependence measurements showed that the acceptor resonance came principally from an emission at 950 nm while the D$_1$ and D$_2$ resonances extended over the whole emission. Time resolved emission measurements showed the existence of two bands, one at 950 nm and the other at 870 nm. The latter decayed rapidly at first with a time constant of the order of 10 nsec and both emissions showed non-exponential decays of approximately 2 μsec. The 950 nm band also showed a shift of the emission peak position with decay and is consistent with a donor-acceptor emission where the centres involved are the D$_1$ and acceptor centres. The high energy emission is either excitonic like or a recombination between an electron on a D_1 centre and a hole in the valence band edge. The D_2 centre is believed to be a non-radiative recombination centre such as a dangling bond in a void.

CONCLUSIONS

ODMR has become an important technique for investigating recombination processes in both crystalline and amorphous semiconductors. A wide range of investigations are under way in the case of III-V and II-VI crystals with interest in vacancy centres, deep donors, implanted ions, transition metal impurities and other deep traps. Studies of amorphous Si and chalcogenide glasses (48) are continuing in parallel with investigations of analogous defects in the crystalline materials. Exciton recombination in layered compounds is being investigated in GaS_xSe_{1-x} and in materials such as HgI_2 and PbI_2.

It is important to note that the detection of resonances via the luminescence can often be linked to other spin dependent processes in semiconductors such as conductivity, Hall effect, photoconductivity (49) and solar cells (50).

ACKNOWLEDGEMENTS - I wish to thank Dr. J.J.Davies, Dr. J.E.Nicholls, Mr. P.Dawson, Mr. N.Killoran, Dr. P.J.Dean and Mr. P.Smith with whom these investigations have been carried out. The collaborations with Professor K.Morigaki from the ISSP, Tokyo and with Dr. M.Gal from the Technical Institute, Budapest have been very stimulating and to them I am very grateful. I am grateful for the generous support of the Royal Society and the Science Research Council.

REFERENCES

1. U.Kaufmann, J.Schneider and A.Rauber, Appl.Phys.Letters 29, 312 (1976)
2. T.A.Kennedy and N.D.Wilsey, Phys.Rev.Letters 41, 977 (1978)
3. J.J.Krebs and G.H.Stauss, Phys.Rev. B16, 971 (1977)
4. A.Abragam and B.Bleaney, Electron Paramagnetic Resonance of Transition Ions (Oxford University Press, Oxford 1970)
5. S.Geschwind, Paramagnetic Resonance, ed. by S.Geschwind, Chapter 5. Plenum, New York (1972)
6. B.C.Cavenett, Proc. Luminescence Conf. Paris, 1978. J.Luminescence 18/19, 846 (1979)
7. B.C.Cavenett, Proc. Int. Conf. Microwave Diagnostics of Semiconductors, Porvoo 1977, ed. by R.Paananen, Helsingfors, Helsinki, p.27.
8. B.C.Cavenett, Luminescence Spectroscopy, ed. by M.D.Lumb (Academic Press, London 1978) Chapter 5.

9. P.Dawson, K.Morigaki and B.C.Cavenett, Proc.Int.Conf.Semicon-
ductors, Edinburgh 1978, ed. by B.H.L.Wilson (Institute of
Physics 1979) p.1023.
10. K.Morigaki, P.Dawson and B.C.Cavenett, Solid State Commun. 28,
829 (1979)
11. P.L.Schmid, J.P.Voitchovsky and A.Mercier, Phys. Stat. Solidi a
21, 443-50 (1974)
12. A.Mercier and J.P.Voitchovsky, Phys.Chem.Solids 36, 1411 (1975)
13. B.C.Cavenett, P.Dawson and K.Morigaki, J.Phys.C. 12, L197 (1979)
14. P.Dawson, B.C.Cavenett and N.Killoran, Solid State Commun. (to
be published)
15. R.F.Brunwin, B.C.Cavenett, J.J.Davies and J.E.Nicholls, Solid
State Commun. 18, 1283 (1976)
16. D.J.Dunstan, B.C.Cavenett, P.Dawson and J.E.Nicholls, J.Phys.C.
1978 (to be published)
17. J.R.James, B.C.Cavenett, J.E.Nicholls, J.J.Davies and D.J.Dunstan,
J.Luminescence 12/13, 447 (1976)
18. J.R.James, J.E.Nicholls, B.C.Cavenett, J.J.Davies and D.J.Dunstan,
Solid State Commun. 17, 969 (1975)
19. J.E.Nicholls, J.J.Davies, B.C.Cavenett, J.R.James and D.J.Dunstan,
J.Phys.C. 12, 361 (1979)
20. D.Block, R.T.Cox, A.Herve, R.Picard, C.Santier and R.Helbig,
Proc. Colloquie Ampere, Dublin, 1977, p.439
21. R.T.Cox, D.Block, A.Herve, R.Picard, C.Santier and R.Helbig,
Solid State Commun. 25, 77 (1978)
22. P.Dawson, B.C.Cavenett and G.Sowersby, Proc.Int.Conf.Recombina-
tion Radiation, Southampton, Solid State Electronics 21, 1451
(1978)
23. P.Dawson and B.C.Cavenett, Proc. Int. Luminescence Conf., Paris,
1978. J.Luminescence 18/19, 853 (1979)
24. P.Dawson and B.C.Cavenett (to be published)
25. C.H.Henry, P.F.Dean and J.D.Cuthbert, Phys.Rev. 166, 754 (1968)
26. R.L.Hollis, J.F.Ryan, D.J.Toms and J.F.Scott, Phys.Rev.Letters
31, 1004 (1973)
27. J.F.Scott and R.L.Hollis, Solid State Commun. 20, 1125 (1976)
28. H.Venghaus, P.E.Simmonds, J.Lagois, P.J.Dean and D.Bimberg,
Solid State Commun. 24, 5 (1977)
29. P.J.Dean, H.Venghaus, J.C.Pfister, B.Schaub and J.Marine,
J.Luminescence 16, 363 (1978)
30. N.Killoran, B.C.Cavenett and P.J.Dean (to be published)
31. D.J.Dunstan, B.C.Cavenett, R.F.Brunwin and J.E.Nicholls, J.Phys.
C. 10, L361 (1977)
32. D.J.Dunstan, J.E.Nicholls, B.C.Cavenett, J.A.Davies and K.V.Reddy,
Solid State Commun. 24, 677 (1977)
33. G.D.Watkins, Phys. Rev. Lett. 33, 223 (1974)
34. J.E.Nicholls, D.J.Dunstan and J.J.Davies, Semicond.Insulators 4.
35. H.G.Grimmeiss, C.Ovren, W.Ludwig and R.Mach, J.Appl.Phys. 48,
5122 (1977)
36. D.J.Dunstan, J.E.Nicholls, B.C.Cavenett, J.J.Davies and K.V.Reddy,
J.Phys.C. (to be published)
37. J.J.Davies and J.E.Nicholls, J.Luminescence 18/19, 322 (1979)
38. J.E.Nicholls and J.J.Davies, J.Phys.C. 12, 1917 (1979)
39. P.J.Dean, C.H.Henry and C.J.Frosch, Phys.Rev. 168, 812 (1968)
40. P.J.Dean and C.H.Henry, Phys.Rev. 176, 928 (1968)
41. H.Kukimoto, C.Henry and F.R.Merritt, Phys.Rev. B7, 2486)1973)
42. C.H.Henry, H.Kukimoto, G.L.Miller and F.R.Merritt, Phys.Rev. B7,
2499 (1973)
43. H.G.Grimmeiss, C.A.Lebebo, C.Ovren and T.N.Morgan, Proc. 12th
Int. Conf. Semiconductors, ed. by M.H.Pilkuhn (Teubner, Stuttgart,
1974)
44. T.N.Morgan, J.Electron Mat. 4, 1029 (1975)
45. T.N.Morgan, Phys.Rev.Lett. 40, 190 (1978)

46. M.Gal, B.C.Cavenett and P.Smith, Phys.Rev.Letters (to be pub-
    lished)
47. K.Morigaki, D.J.Dunstan, B.C.Cavenett, P.Dawson and J.E.Nicholls,
    Solid State Commun. 26, 981 (1978)
48. H.Suzuki, K.Murayama and T.Ninomiya, J.Phys.Soc. Japan 46, 693
    (1979)

DEEP LEVEL SPECTROSCOPY IN SEMICONDUCTORS

BY OPTICAL EXCITATION

H.G. Grimmeiss
Lund Institute of Technology
Department of Solid State Physics
Box 725, S-220 07 LUND 7, Sweden

1. INTRODUCTION

Replacing an atom of the host-lattice by a foreign atom results in lattice defects with physical properties which depend considerably on the particular atom introduced. Such lattice defects are well understood, in many cases, for instance, in Ge and Si when the foreign atom belongs to one of the adjacent groups in the periodic table. The potential binding the extra electron or hole at the impurity atom can then often be approximated by a hydrogen-like potential which introduces a localized energy level in the otherwise forbidden energy gap for the impurity ground state and, in addition, a series of excited states. The ground state energy of such centers is typically less than 100 meV in most semiconductors. They are therefore called "shallow" impurity levels, because they lie close to one of the energy band edges. Shallow impurities are widely used in semiconductor technology for modifying the degree and type of electrical conductivity.

Impurities which create energy levels further away from the band edge are called "deep" impurity levels. In Si and Ge impurities not belonging to one of the adjacent groups in the periodic table, often form such deep energy levels. But not only impurities create deep energy levels in semiconductors. Very often native defects have binding energies which are much larger than those for shallow impurities.

In this paper the electronic properties of "deep" energy levels are discussed. The emphasis will be on isolated point defects. Such defects are often assumed to be single substitutional impurities, even though it may turn out that this picture is too simple. Recently increasing evidence has been accumulated showing that many impurities form some kind of complex either with native defects or other impurities. Knowledge of the nature of the defect is of utmost importance for succesful theoretical models.

Deep energy levels seem to be present in all known semiconductors. One of their most important properties is the ability to control the carrier lifetime even when present in small concentrations. This is readily shown by Shockley-Read-Hall statistics [1,2]. According to these statistics, the lifetime τ of excess charge carriers in a semi-conductor with a single energy level is given by

$$\tau = \frac{c_p(p_o + p_1) + c_n(n_o + n_1)}{c_n c_p N_{TT}(n_o + p_o)} \tag{1}$$

where c_n and c_p are the average values of the capture constants of electrons and holes over the states in the bands, n_o and p_o are the free carrier concentrations in thermal equilibrium, and n_1 and p_1 are the electron and hole concentrations when the Fermi level E_F falls at E_T, the energy position of the energy level. N_{TT} is the number of centers per unit volume. Let us now, for the sake of illustration, con-sider an n-type semiconductor ($E_F > E_{iF}$ where E_{iF} is the intrinsic Fermi level) with an energy level above the Fermi level ($E_T > E_F$). In this case, we have

$$n_1 \gg n_o \gg p_o \gg p_1 \tag{2}$$

a condition, for example, which often exists in the effective space charge region of a Schottky barrier. The above equation then reduces to

$$\tau = \frac{n_1}{N_{TT} c_p n_o} = \frac{1}{c_p n_o} \cdot \frac{N_c}{N_{TT}} \exp\left(-\frac{E_c - E_T}{kT}\right) \tag{3}$$

showing the significant influence on the excess carrier lifetime not only of a large value of $E_c - E_T$ but also of a large capture constant. Deep energy levels are therefore very important for the fabrication of devices.

It is quite obvious for the reasons just mentioned that the energy position and the capture constant are important parameters for the caracterization of deep energy levels. These parameters are in turn related to the thermal emission rate e^t by the detailed balance relationship

$$e_n^t = c_n N_c \exp(-\Delta G_n / kT) \tag{4}$$

$$e_p^t = c_p N_v \exp(-\Delta G_p (kT) \tag{5}$$

It should be noted that when the Gibbs free energy G is used instead of the enthalpy, the above equations do not include the electronic de-generacy factors in the usual way, as shown by Engström and Alm [3].

ΔG is then the free energy (Gibbs's free energy) needed to excite a charge carrier from an energy level into the nearest energy band.

Hence, the electronic properties of deep energy levels are often characterized by their capture constants, emission rates and energy position. Optical emission rates e^o in their turn are correlated to the photoionization cross section σ^o by the relation

$$e^o = \sigma^o \phi \tag{6}$$

where ϕ is the photon flux of the incident light for measuring e^o. Capture cross sections are easily calculated by dividing the capture constant by the thermal velocity.

The purpose of this paper is to discuss measuring techniques for the determination of emission rates. The emphasis will be on optical emission rates. Information on the energy position of a deep center may be obtained from the spectral distribution of optical emission rates. Techniques for measuring capture constants will be presented in the next paper.

2. MEASURING TECHNIQUES

Emission rates are most commonly measured by junction space charge techniques. Both steady-state and transient methods are employed. A number of different techniques have been developed during the last few years and it would be far beyond the scope of this paper to cover even the most common of them. The selection of methods described is therefore to some extent arbitrary.

2.1 Dual-light-source steady-state photocurrent method

Let us start with the dual-light-source spectroscopy technique (DLSS). The current generated in a reverse biased junction consists of two components, the electron generation current J_n and the hole generation current J_p. Hence, the current density of a reverse-biased junction can be calculated according to

$$J_R = J_n + J_p = q \int_{x_1}^{x_2} \left[\frac{1}{D} U_n + (1 - \frac{1}{D}) U_p \right] dx \tag{7}$$

where $x_2 - x_1 = W - W_o$ is the effective generation region width (Fig. 1) [4]. U_n and U_p are the total net rates of electron and hole emission, respectively, and D is a factor taking into account displacement currents [5,6]. Because capture processes can be neglected in reverse-

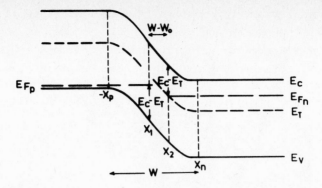

Fig. 1. Band diagram of a p-n junction with deep energy levels at E_T

-biased junctions, the net rate of electron and hole emission is given by $U_n = e_n n_T$ and $U_p = e_p p_T$, where n_T and p_T are the concentrations of centres occupied by electrons and holes, respectively. If the center has only two charge states, then obviously

$$n_T + p_T = N_{TT}. \tag{8}$$

Eq. (7) can therefore be rewritten as

$$J_R = q(W - W_o) \left\{ \frac{1}{D} e_n n_T + (1 - \frac{1}{D}) e_p (N_{TT} - n_T) \right\} \tag{9}$$

It should be realised that the emission rates in Eq. (9) are composed of the sum of the thermal and optical emission rates $e^t + e^o$ and that at steady state ($U_n = U_p$) we have

$$n_T(\infty) = \frac{e_p}{e_n + e_p} \tag{10}$$

Below the freeze-out temperature ($e^o \gg e^t$) one therefore obtains for the steady-state photocurrent density

$$J_R^o(\infty) = q(W - W_o) N_{TT} \frac{e_n^o e_p^o}{e_n^o + e_p^o} \tag{11}$$

Any current through a reverse-biased junction is a generation current for which recombination processes can be neglected and is therefore very suitable for the investigations of emission rates. However, it is quite clear from Eq. (11) that the spectral distributions of e_n^o and e_p^o cannot be investigated separately from steady-state currents using one light source. The reason for that is that the occupancy of the centers is changed when the photon energy is varied. A constant occupancy is, however, easily achieved by using a second light source with properly chosen constant photon energy $h\nu_s$. The steady-state photocurrent density is then given by

$$J^O_{R\ell}(\infty) = q(W-W_O)N_{TT} \frac{(e^O_n + e^O_{ns})(e^O_p + e^O_{ps})}{e^O_n + e^O_{ns} + e^O_p + e^O_{ps}} \qquad (12)$$

where e^O_{ns} and e^O_{ps} are the optical emission rates due to the second light source. The spectral distribution of e^O_n and e^O_p can now be separately investigated depending on the photon energy $h\nu_s$ and photon flux ϕ_s of the second light source. If $h\nu_s$ and ϕ_s are chosen such that $E_c-E_T < h\nu_s < E_T-E_v$ (i.e. $e^O_{ps} = 0$) and $e^O_{ns} \gg e^O_n + e^O_p$ (i.e. $\phi_s \gg \phi$), then the steady-state photocurrent density is

$$J^O_{Rh}(\infty) = q(\bar{W}-W_O)N_{TT}e^O_p \qquad (13)$$

for a single impurity level in the upper half of the bandgap (Fig. 2).

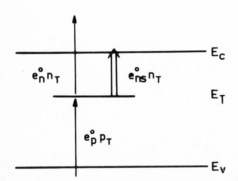

Fig. 2. Generation of photocurrent J^O_{Rh} due to simultaneous illumination with two light sources of photon energies $E_c-E_T < h\nu_s < E_T-E_v$ and $E_T-E_v < h\nu < E_g$, respectively.

Because during the measurement the occupancy and, hence, $W-W_O$ is kept constant, J^O_{Rh} is proportional to e^O_p for all photon energies $h\nu$ smaller than the bandgap E_g. Such a measurement is readily performed and not time consuming. For the sake of illustration we have chosen a gold-related energy level in the upper half of the bandgap in Si to demonstrate the measuring technique. The spectral distribution of e^O_p of this particular center is easily obtained over about three orders of magnitude by plotting J^O_{Rh} as a function of photon energy without any further analysis (Fig. 3).

The spectral distribution of e^O_n is obtained by choosing $h\nu_s$ and ϕ_s such that $E_T-E_v < h\nu_s < E_g$ and both e^O_{ns} and e^O_{ps} are much larger than e^O_n and e^O_p for variable photon energies (Fig. 4). It has been shown [7] that the increase in the steady-state photocurrent density due to illumination by the second light source $\Delta J^O_R = J^O_{R\ell}(\infty) - J^O_R(\infty)$ can then

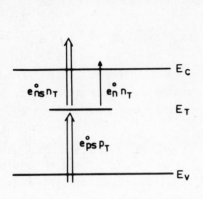

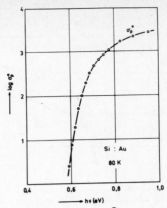

Fig. 3. Logarithm of σ_p^o versus photon energy for a gold-related center in silicon as obtained from a measurement of J_{Rh}^o.

Fig. 4. Generation of photocurrent ΔJ_R^o due to simultaneous illumination with two light sources of photon energies $E_T-E_V < h\nu_s < E_g$ and $E_c-E_T < h\nu < E_T-E_V$, respectively.

be expressed as

$$\Delta J_R^o = q(W-W_o)N_{TT}\left[(1-b)^2 e_p^o + b^2 e_n^o\right]$$

$$= (1-b)^2 J_{Rh}^o + q(W-W_o)N_{TT}b^2 e_n^o \tag{14}$$

where

$$b = \frac{n_T}{N_{TT}} = \frac{e_{ps}^o}{e_{ns}^o + e_{ps}^o} = 1 - \frac{J_R^o}{J_{Rh}^o} \tag{15}$$

is the electron occupancy of the energy level when the junction is illuminated with photons of energy $h\nu_s$ alone (see Eq. (10)). Because $\phi_s \gg \phi$, b can be considered as constant during the measurements and, hence, a plot of $\Delta J_R^o - (1-b)^2 J_{Rh}^o$ versus $h\nu$ gives the spectral distribution of e_n^o for all photon energies $h\nu$ smaller than the bandgap (Fig. 5).

Very often one is only interested in a limited energy region of the spectrum close to the threshold energy. The measurements can then be appreciably simplified by choosing $h\nu < E_T-E_V$. For these photon

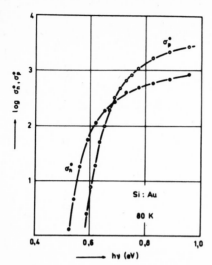

Fig. 5. Logarithm of σ_n^o versus photon energy for a gold related center in silicon as obtained from a measurement of ΔJ_R^o. For comparison σ_p^o is also plotted.

energies the optical emission rate e_p^o vanishes and $J_{Rh}^o = 0$. Eq.(14) reduces then to

$$\Delta J_R^o = q(W-W_o)N_{TT}b^2e_n^o \tag{16}$$

and the increase in the steady--state photocurrent density is directly proportional to e_n^o.

2.2 Transient photocurrent technique

Although e_n^o and e_p^o can be calculated from Eqs(13) and(16) if $\overline{W}-W_o$, N_{TT} and b are known, absolute values of emission rates are most commonly measured directly using transient techniques. In general, it is sufficient to determine the absolute value of one of the emission rates at a particular photon energy. Knowning the spectral distributions of J_R^o, ΔJ_R^o and J_{Rh}^o, absolute values of e_n^o and e_p^o can then be calculated for all photon energies smaller than the bandgap, using Eqs(13),(14) and (15). For an energy level in the upper half of the bandgap it is most convenient to measure the absolute value of e_n^o for a photon energy $h\nu < E_T-E_v$ (i.e. $e_p^o = 0$). Using the fact that $dn_T/dt = U_p-U_n$ we obtain for the time dependence of occupied energy levels within the effective generation region of the junction

$$\frac{dn_T}{dt} = (e_p^o + e_p^t)(N_{TT} - n_T) - (e_n^o + e_n^t)n_T. \tag{17}$$

Below the freeze-out temperature ($e_n^t = e_p^t = 0$) integration of Eq.(17) gives

$$n_T(t) = \frac{e_p^o}{e_n^o+e_p^o}N_{TT} + \left[n_T(o) - \frac{e_p^o}{e_n^o+e_p^o}N_{TT}\right]\exp\left[-(e_n^o+e_p^o)t\right] \tag{18}$$

which for $e_p^o = 0$ reduces to

$$n_T(t) = n_T(o)\exp(-e_n^ot). \tag{19}$$

The photocurrent transient which is observed due to the illumination with photons of energy $h\nu < E_T-E_v$ may be calculated for temperatures below the freeze-out temperature ($e^t = 0$) by inserting Eq.(19) in Eq.(9)

$$J_R^o(t) = q(W-W_o)\frac{1}{D}\,e_n^o n_T(o)\exp(-e_n^o t) = J_R^o(o)\exp(-e_n^o t). \tag{20}$$

This result shows that independently of the initial conditions - as long as $n_T(o) \neq 0$ - the optical emission rate e_n^o is readily obtained from the time constant of the photocurrent transient. Using Eq.(6) the absolute value of the photoionization cross section σ_n^o can be calculated from the optical emission rate if the photon flux ϕ of the incident light for generating the transient is known. Hence, from one transient measurement together with the steady-state currents J_R^o, ΔJ_R^o and J_{Rh}^o the spectra of absolute photoionization cross sections are obtained.

The DLSS technique is most conveniently employed with p-n junctions. If Schottky barriers only are available (as for example in some of the II-VI compounds) the energy region of the spectrum which can easily be investigated is then limited by the barrier height. With Schottky diodes emission rates are therefore often studied by capacitance techniques.

2.3 Constant capacitance technique

The most commonly used methods with capacitance techniques are transient measurements at constant reverse bias [8]. In order to obtain simple exponential time dependences transient measurements are often performed in samples with moderate deep-level concentrations such that the concentration of deep energy levels N_{TT} is much smaller than the concentration of shallow levels N_D. This implies that rather small capacitance changes are obtained during the measurements. However, the sensitivity of the measurements can be increased considerably by using highly compensated samples and measuring the change in reverse bias at constant barrier capacitance [8a] instead of monitoring the change in capacitance at constant reverse bias.

The principles of the technique have been described in detail by Grimmeiss, Ovrén and Mach [9,10] and are outlined in Fig. 6 and 7. The energy band diagram (Fig. 6a) represents a Schottky barrier on an n-type semiconductor with an acceptor-like deep impurity ($N_{TT} < N_D$). In the region, W_o, nearest to the neutral material of the diode the deep energy levels at E_T are filled with electrons. In the remaining part of the space-charge region, $W-W_o$, the concentration of deep energy levels, n_T, which are occupied by electrons is determined by the

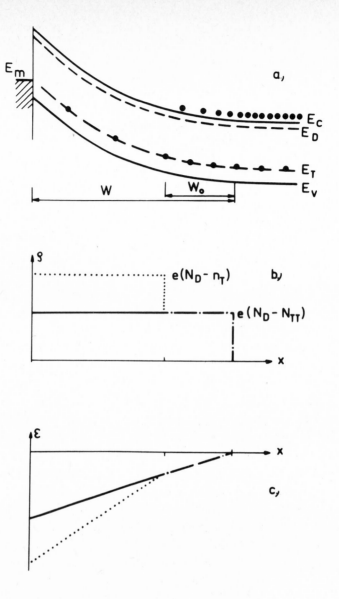

Fig. 6. a) Energy diagram of a Schottky barrier with N_D shallow donors at energy E_D and N_{TT} deep energy levels at energy E_T.

b) Distribution of the space-charge density ϱ when all deep centers are occupied with electrons (——) and when some are empty (...).

c) Electric field distribution in the space-charge region when all deep centers are occupied with electrons (——) and when some are empty (...).

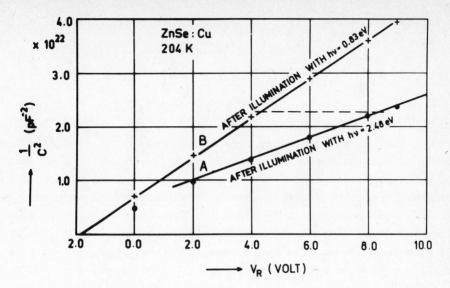

Fig. 7. Plots of C^{-2} versus voltage when $n_T = 0.5\ N_{TT}$ (curve A) and $n_T = N_{TT}$ (curve B), respectively.

emission rates and, hence, $0 < n_T < N_{TT}$. The distribution of the charge density ϱ in the space-charge region of the barrier depends, however, on the occupancy of the energy levels. Fig. 6b shows the distribution of the space-charge density when all deep centers are filled with electrons (solid curve) and when some of them are empty (dotted curve). The electric field distribution corresponding to the two charge distributions can be calculated from Poisson's equation giving

$$\mathcal{E}(x) = q/\mathcal{E}\mathcal{E}_0\ [\,(N_D - n_T)x + N_{TT}W_0 - N_D W + n_T(W - W_D)\,] \tag{21}$$

for $0 < x < W - W_0$ and

$$\mathcal{E}(x) = q/\mathcal{E}\mathcal{E}_0\,(N_D - N_{TT})\,(x - W) \tag{22}$$

for $W - W_0 < x < W$ and an arbitrary value of n_T (Fig. 6c, dotted curve).

For $n_T = N_{TT}$ Eq. (22) is valid in the total space charge region $0 < x < W$ (Fig. 6c, solid curve). From this field distribution the capacitance-voltage curve (Fig. 7) is expected to give a straight line only when all deep centers are filled with electrons. If some of them are empty a straight line is obtained only when the reverse bias is large enough that the contribution from the region W_0 can be neglected.

Integrating the electric field gives the voltage drop across the barrier,

$$V-V_R = \int\limits_{0}^{W-W_0} \mathcal{E}(x)\,dx + \int\limits_{W-W_0}^{W} \mathcal{E}(x)\,dx = I_1 + I_2 \tag{23}$$

where V_R is the reverse bias voltage. Eqs.(21),(22) and (23) clearly show that the two integrals may be written as $I_1 = c_0 n_T + c_1$ and $I_2 = c_2$, since W_0 is independent of the reverse bias [4] if during the measurements the diode capacitance and, hence, the total width of the depletion region, W, is kept constant. However, this implies that when the occupancy of the deep center is changed, the change in voltage ΔV needed to keep the diode capacitance constant is proportional to n_T. In heavily compensated samples the change in voltage ΔV due to the recharging of the deep energy levels may be rather large. According to Fig. 7, ΔV is of the same order as the total voltage over the diode, depending on which capacitance value is kept constant. The data presented in Fig. 7 are taken from measurements on Cu-doped ZnSe [10]. Curve A of Fig. 7 is obtained with $n_T = (1/2)N_{TT}$ and curve B with all deep centers occupied by electrons. Hence, by changing the occupancy of the center due to illumination with photons of energy $h\nu$, the optical emission rate in Schottky diodes is readily obtained by monitoring the change in voltage ΔV needed to keep the diode capacitance constant. Because ΔV is proportional to n_T the time constant of the signal is given by Eq.(18) which for $e_n^o = 0(h\nu < E_c - E_T)$ directly gives the photoionization cross section of holes σ_p^o in absolute values if the photon flux of the incident light is known.

The constant capacitance spectroscopy (CCS) technique has made it possible to perform measurements with higher accuracy and in a larger temperature region than in previous investigations. As an example, data are presented in Fig. 8 which have been obtained for a Cu-related acceptor level in ZnSe at different temperatures [9,10].

By a similar analysis the same technique may be used to measure the spectral distribution of σ_n^o [11]. If all the deep energy levels in the lower part of the bandgap are initially filled with electrons $(n_T(o) = N_{TT})$ and the diode is illuminated with photons of energy $h\nu \geq E_c - E_T$ the total change in occupancy is then according to Eq.(18) given by

$$n_T(o) - n_T(\infty) = \frac{e_n^o}{e_n^o + e_p^o}\, N_{TT} = B\,\Delta V \tag{24}$$

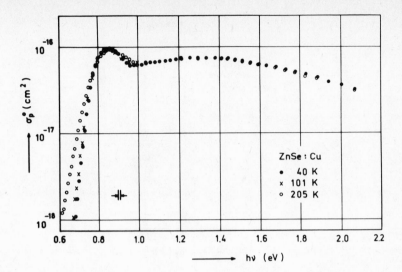

Fig. 3.   Spectra of the optical cross section σ_p^O in ZnSe:Cu at different temperatures obtained with the CCS technique.

The time constant τ of the change in voltage ΔV due to the illumination is

$$\tau = \left[(e_n^O + e_p^O) \right]^{-1} \tag{25}$$

Because ΔV is proportional to the change in occupancy, the photoionization cross section of electrons σ_n^O is obtained by combining Eq.(24) with (25) as

$$\sigma_n^O = \frac{B \, \Delta V}{\tau \, \phi} \tag{26}$$

where B is constant for constant photon flux. Data obtained with this CCS technique for the copper-related center in ZnSe [9,10] are shown in Fig. 9 as an example. Again, from the spectral distribution of the photoionization cross sections σ_n^O and σ_p^O information on the energy position E_T of the center may be obtained.

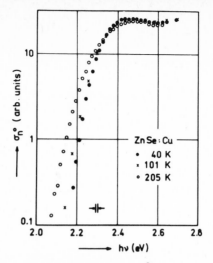

Fig. 9. Spectra of the optical cross section σ_n^o in ZnSe:Cu at different temperatures obtained with the CCS technique.

2.4 Steady-state constant capacitance method

Transient measurements are often more time consuming than steady-
state measurements. Using two independent light-sources [12] as in the
case of the steady-state photocurrent method, a steady-state constant
capacitance spectroscopy (SCCS) technique can be employed for heavily
compensated samples. Because the SCCS technique is also readily applied
to Schottky diodes this method may therefore be used with great advantage in II-VI compounds, where compensation is often easily
achieved.

For the sake of illustration it is assumed that the level concerned is located in the lower half of the bandgap. If the capacitance
is again kept constant, the change in voltage is proportional to the
change in occupancy. Let us start by illuminating the diode with a
light source of intensity ϕ_s and constant energy $h\nu_s$ such that
$E_T-E_V < h\nu_s < E_c-E_T$ $(e_{ns}^o = 0)$. The concentration n_{TS} of deep energy
levels which are filled with electrons is then equal to N_{TT}. Additional
illumination of the sample with a second light source of variable
energy $h\nu$ such that $h\nu > E_c-E_T$ (Fig. 10) changes the concentration
of filled centers to $N_{TT}(e_p^o + e_{ps}^o)(e_n^o + e_p^o + e_{ps}^o)$. The change in concentration Δn_T due to the illumination of the sample with photons of
variable energy $h\nu$ is given by

$$\Delta n_T = N_{TT} - \frac{e_p^o + e_{ps}^o}{e_n^o + e_p^o + e_{ps}^o} N_{TT} = \frac{e_n^o}{e_n^o + e_p^o + e_{ps}^o} N_{TT} \qquad (27)$$

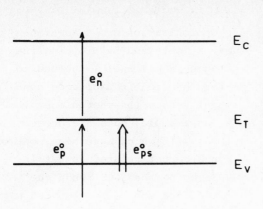

E_c

e_n^o

E_T

e_p^o e_{ps}^o

E_v

Fig. 10. Energy and intensity conditions for measuring spectra of e_n^o using the steady-state constant capacitance method.

If the intenstiy ϕ_s is chosen so that e_{ps}^o is much larger than both e_n^o and e_p^o, Eq.(27) reduces to $\Delta n_T \approx e_n^o N_{TT}/e_{ps}^o$. The change in voltage ΔV needed to keep the diode capacitance constant during the measurements is therefore proportional to e_n^o and the spectral distribution of the photoionization cross section σ_n^o is obtained by plotting ΔV as a function of the photon energy $h\nu$.

This measuring technique is based on the one hand on the assumption that the occupancy is kept constant during the measurements and on the other hand on the fact that the signal obtained is due to the change of the occupancy. The error in the measurement is therefore of the order of e_n^o/e_{ps}^o and, for example, less than 1 ^{o}/oo if $e_{ps}^o > 10^3 e_n^o$.

We next turn to an outline of the principles for measuring the spectrum of the photoionization cross section of holes $e_p^{o/}$. The initial condition for the experiment is established by illuminating the sample with photons of constant energy $h\nu_s$ such that $E_g > h\nu_s > E_c-E_T$ (Fig. 11). The photon flux of the light source is ϕ_s. The concentration of filled centers is then according to Eq.(10) given by $e_{ps}^o N_{TT}/(e_{ns}^o+e_{ps}^o)$. Additional illumination of the sample with photons of variable energy $h\nu$ such that $h\nu < E_c-E_T$ ($e_n^o = 0$) changes the concentration of filled centers to $N_{TT}(e_p^o + e_{ps}^o)/(e_{ns}^o + e_p^o + e_{ps}^o)$. The change in concentration Δn_T due to the illumination with photons of variable energy $h\nu$ is then given by

$$\Delta n_T = \frac{e_{ps}^o}{e_{ns}^o + e_{ps}^o} N_{TT} - \frac{e_p^o + e_{ps}^o}{e_{ns}^o + e_p^o + e_{ps}^o} N_{TT} \qquad (28)$$

Choosing ϕ_s such that both e_{ns}^o and e_{ps}^o are much larger than e_p^o, the right hand side of Eq.(28) reduces to approximately $N_{TT}e_p^o e_{ns}^o/(e_{ns}^o+e_{ps}^o)^2$, showing that ΔV is proportional to e_p^o. The error in the measurements is of the order of $e_p^o/(e_{ns}^o + e_{ps}^o)$ and can be neglected if ϕ_s is made large.

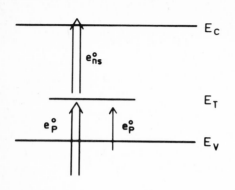

Fig. 11. Energy and intensity conditions for measuring spectra of e_p^o using the steady-state constant capacitance method.

Hence, with the above conditions the spectral distribution of the photoionization cross section e_p^o may be obtained by plotting the change in voltage ΔV needed to keep the diode capacitance constant versus the photon energy $h\nu$. As an example, data are presented in Fig. 12 which have been obtained for a copper related center in ZnSe using the SCCS technique [13]. These data should be compared with the results shown in Fig. 8. From the spectra of e_n^o and e_p^o information about the energy position of the center is obtained.

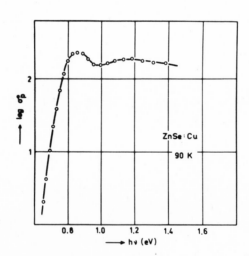

Fig. 12. Spectrum of the optical cross section σ_p^o in ZnSe:Cu obtained with the steady state constant capacitance technique.

2.5 Constant photoconductivity method

Although junction space charge techniques are very useful for the investigations of the electronic properties of deep energy levels they nevertheless have one serious drawback: all the excitation processes studied occur in the space charge region of the barrier and, hence, in the presence of high electric fields. This may or may not influence

the measurements. A careful check of all data with respect to any influence of the electric field is therefore always advisable. One way
of checking the data is to perform similar investigations in bulk material. For radiative centers a number of photoluminescence techniques
have been developed [14,15]. If the defect is non-radiative, photoconductivity measurements do not always give direct information on photoionization cross sections, which are required for the determination of
the energy position of the center and for comparison with data obtained
by other measuring techniques [6,16]. The reason for this is that photoconductivity often depends in a complicated manner on capture constants and emission rates. This is readily shown by considering, for
example, an n-type semiconductor with a single energy level at E_T in
the upper half of the bandgap and concentration N_{TT}. Illuminating the
sample with photons of energy $h\nu < E_T-E_V$ and choosing a photon flux ϕ
such that the excess electron concentration is much larger than the
electron concentration in thermal equilibrium, the free electron concentration n is given by the density of empty centers and, hence, according to Eq.(8) we have [16]

$$n \cong N_{TT} - n_T. \tag{29}$$

The rate equation describing the excitation and recombination processes between the energy level and the conduction band

$$\frac{dn}{dt} = e_n^o n_T - c_n n(N_{TT} - n_T) \tag{30}$$

reduces then at steady state to

$$e_n^o(N_{TT} - n) - c_n n^2 = 0 \tag{31}$$

where c_n is the total capture constant for electrons. Solving Eq.(31),
the free electron concentration can be expressed as

$$n = - \frac{e_n^o}{2c_n} + \sqrt{\left(\frac{e_n^o}{2c_n}\right)^2 + N_{TT}\frac{e_n^o}{c_n}}. \tag{32}$$

Because the photoconductivity of a sample is proportional to n, Eq.(32)
implies that it may be difficult to obtain the spectral distribution of
optical emission rates from photoconductivity measurements under these
conditions unless c_n and $\bar{N}_{TT}$ are known, which usually is not the case.
An additional important aspect of photoconductivity measurements is
that the response and decay times of the signal in samples containing

deep energy levels are often large due to recharging of the centers. Decay times of several hours are not unusual and have, for example, been observed in semiinsulating GaAs:O [17,18]. Eq.(32) also shows that the intensity dependence of the photoconductivity may be rather complex. Sometimes the measuring conditions can be improved by using small excitation densities, such that $n \ll N_{TT}$. Eq.(32) reduces then to

$$n = \sqrt{N_{TT} \frac{e_n^o}{c_n}} = \sqrt{\phi N_{TT} \frac{\sigma_n^o}{c_n}} \tag{33}$$

and the photoconductivity signal depends on the square root of the photon flux. However, the sample often contains several deep energy levels and $n \sim (e_n^o)^f$ where $0 < f < \infty$. The exponent f may vary with photon energy, photon flux and/or temperature, which makes proper corrections difficult in many cases [16].

Most of these difficulties can be avoided by using the constant photoconductivity method. The important feature of this technique is the fact that the photon flux of the incident monochromatic light is adjusted so that the photoconductivity signal is kept constant when the photon energy changes. Keeping the photoconductivity constant means in the above example that the occupancy of the energy levels involved is unchanged and, hence, that both n and n_T are constant. From Eq.(31) it then follows that e_n^o is constant, and combining with Eq.(6) we obtain

$$\sigma_n^o(h\nu) = B' \frac{1}{\phi(h\nu)} \tag{34}$$

where B' is independent of photon energy. Hence, by plotting the inverse of the photon flux at a constant photoconductivity signal against the photon energy $h\nu$, the spectrum of the photoionization cross section σ_n^o is obtained [16].

For the sake of illustration preliminary data are presented in Fig. 13 which have been measured in gold doped silicon using the constant photoconductivity method [19]. These data are to be compared with the results obtained with junction space charge techniques shown in Fig. 5. Although further experiments have to be performed, the data presented in Fig. 13 may indicate that the photoionization cross section spectra of gold-doped silicon obtained with photocurrent and photoconductivity methods are not the same. Previous measurements on oxygen-doped GaAs showed no difference in spectra when junction space charge techniques and/or bulk measurements were used [16].

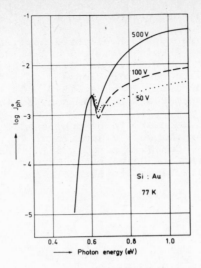

Fig. 13. Photoconductivity spectra of gold doped silicon at different biases.

3. CONCLUSION

Difficulties which arise in the analysis of deep-level data are seldom caused by the lack of adequate measuring techniques. During the last few years a number of different techniques have been developed which make it possible to measure the electronic properties of deep energy levels not only accurately but also in almost all kinds of samples. It has already been mentioned that the selection of measuring techniques described in this paper is somewhat arbitrary. Many other interesting methods have been suggested and successfully applied. Further examples will be discussed in the next paper.

REFERENCES

1. W.Shockley and W.T.Read, Phys.Rev. 87, 835 (1952)
2. R.N.Hall, Phys.Rev. 86, 600 (1952)
3. O.Engström and A.Alm, Solid State Electronics 21, 1571 (1978)
4. S.Braun and H.G.Grimmeiss, J.Appl.Phys. 44, 2789 (1973)
5. O.Engström, Thesis, Lund 1975.
6. H.G.Grimmeiss, Ann.Rev.Mater.Sci. 7, 341 (1977)
7. S.Braun and H.G.Grimmeiss, J.Appl.Phys. 45, 2658 (1974)
8. See for example: D.V.Lang, J.Appl.Phys. 45, 3014 (1974)
8a. G.Goto, S.Yanagisawa, O.Wada and H.Takanashi, Appl.Phys.Lett. 23, 150 (1973);
    J.A.Pals, Solid State Electronics 17, 1139 (1974)
9. H.G.Grimmeiss, C.Ovrén and R.Mach, Int.Conf. on the Phys. of Semicond., 14th. Edinburgh, 1978, Inst.of Phys.Conf.Series No.43,p.273

10. H.G.Grimmeiss, C.Ovrén and R.Mach, to be published in J.Appl. Phys., Sept. 1979.

11. H.G.Grimmeiss, C.Ovrén and J.W.Allen, J.Appl.Phys. 47, 1103 (1976)

12. A.M.White, P.J.Dean and P.Porteous, J.Appl.Phys. 47, 3230 (1976)

13. N.Kullendorff, private communication

14. B.Monemar and L.Samuelson, Phys.Rev. B18, 809 (1978)
    H.G.Grimmeiss, B.Monemar and L.Samuelson, Solid State Electronics 21, 1505 (1978)

15. H.G.Grimmeiss and B.Monemar, to be published

16. H.G.Grimmeiss and L-Å.Lebedo, J.Appl.Phys. 46, 2156 (1975)

17. F.Prat and E.Fortin, Can.J.Phys. 50, 2551 (1972)

18. A.W.Lin, Thesis, Stanford University, Stanford, 1974.

19. H.G.Grimmeiss and J.M.Yanez, to be published

DEPLETED LAYER SPECTROSCOPY

A. Mircea, D. Pons* and S. Makram-Ebeid
Laboratoires d'Electronique et de Physique Appliqueé (L.E.P.)
3 Avenue Descartes, 94450 Limeil-Brévannes, France

1. INTRODUCTION

Depleted Layer Spectroscopy (abreviated DLS) is the general de-signation of a body of experimental techniques, characterized by the following common features:

a) aim to *detect, count* and *characterize* the localized electronic states, due to defects or/and impurities, which have energy levels in the semiconductor band gap;

b) exploit, in order to achieve this program, the specific pro-perties of *depleted layers*, formed by p-n junctions, metal-semicon-ductor barriers, or metal-insulator-semiconductor structures.

The depleted layers obtained due to the existence of an electric field which sweeps the free carriers, electrons as well as holes, away - having behind the fixed space-charge of the ionized defects and impurities. Indeed it is the charge in this space charge that actually constitutes the useful information in DLS experiments, therefore another name - Space Charge Spectroscopy - is also adequate for them.

The basic physical phenomena involved in DLS are:

a) *Optical emission* of electrons or holes, out from the localized states to the conduction/valence bands;

b) *Thermal emission* of same;

c) *Field emission* of same.

Accordingly we distinguish three broad classes of DLS methods; see section 3 below. Out of these three classes, only the first two have received attention until now.

Depleted Layer Spectroscopy is therefore also, essentially, *emission spectroscopy*, in relation to the fact that in a depleted layer there are no carrier capture processes. This is the single most important feature of DLS and is the main reason why these techniques have found an ever increasing domain of application; indeed, the se-

* Formerly with L.E.P. Present address: Laboratoire Central de Recherche (L.C.R.)
Thomson-CSF, Domaine de Corbeville, 91401 Orsay, France

paration of the emission processes from the capture ones greatly simplifies the analysis of the experiments.

Nevertheless the free carrier capture processes can also be studied. This becomes possible since in a p-n junction or Schottky barrier the electric field can be varied at will, by changing the applied bias. Among the different classes of DLS techniques listed below, the ones based upon the analysis of transients are best suited for this kind of study.

Neglecting the free carrier capture processes in the depleted layer is only legitimate if the carrier mobility is large enough so that re-trapping of the emitted carriers is not important. For this reason, the application of DLS techniques to very low mobility semi-conductors (e.g., amorphous silicon) is not immediate and certainly requires particular precautions.

Even in the case of high-mobility semiconductors, the influence of the residual free carriers in the transition region (from depleted to neutral) must be taken into consideration. Section 5.1 below is devoted to some aspects of this question.

Finally let us stress again that the physical quantity which is monitored in DLS is a net variation of electric charge. If the elec-tric charge does not vary, no signal will be present in DLS experiments which means that intra-center electronic transitions (e.g., from the ground state to a localized excited state of the same defect) *cannot*, at least directly, be studied.

From this point of view the designation "Space Charge Spectroscopy" is particularly suggestive. We have, however, preferred "Depleted Layer Spectroscopy" to avoid a possible confusion with space-charge--limited current (SCLC) methods, where the monitored quantity is *mobile* space-charge, and which do not fall within the scope of this survey.

The material which follows can be divided in two part. The first part is tutorial and is not meant to be complete, but merely to comple-ment the clearly existing comprehensive surveys [1-3]. The second part discusses a couple of recent developments in which the author's labora-tory has played an active role.

More emphasis is put on the thermal emission aspects, as compared to optical emission ones since these are discussed in prof Grimmeiss's talk at this conference.

2. PHYSICAL CONCEPTS INVOLVED IN DEPLETED LAYER SPECTROSCOPY

2.1 The kinetic equations

In all forms of DLS analysis the starting point is the differential equation describing the kinetics of the occupation factor f for an individual, localized electronic state (Fig. 1):

$$\frac{df}{dt} = -(e_n + e_n^o + e_p + e_p^o + c_n n + c_p p)f + c_n n + e_p + e_p^o \qquad (2.1)$$

where $0 < f(t) < 1$. n(p) are the free electron (hole) concentrations; $e_n(e_p)$ the thermal electron (hole) emission rates, from the localized

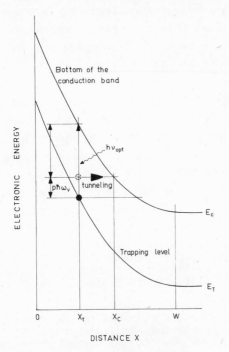

Bottom of the conduction band

$h\nu_{opt}$

$p\hbar\omega_v$ tunneling

E_c

Trapping level

E_T

0   X_t X_c W

DISTANCE X

state to conduction (valence) band; $e_n^o \equiv \sigma_n^o \phi$ $(e_p^o \equiv \sigma_p^o \phi)$ are the emission rates induced by a light flux ϕ; $c_n \equiv \sigma_n v_{thn}$ $(c_p \equiv \sigma_p v_{thp})$ are the free electron (hole) capture coefficients. Note that *field emission* is implicit in $e_n(e_p)$ as well as $e_n^o(e_p^o)$ when (2.1) is applied to a depleted layer.

It is important to remember that a single first-order differential equation suffices to describe all the direct exchanges of carriers between the localized state and the free carrier bands. In experimental practice this means that when the observed signals are characterized by several distinct time constants, they must come from more than one kind of defect.

Fig. 1. Electric potential distribution in a Schottky barrier and the three kinds of bound-to-free electron emission from a deep state to the conduction band.

On the other hand, a continuous distribution of time constants, although it might come from a distribution of states, is more often related to inhomogenities:

a) electric field inhomogenity coupled with a strong contribution of field emission to the total emission rates;

b) variable free carrier concentrations n (or p), typical for the transition region;

c) in the case of optical emission experiments, variable optical emission rates due to strong light absorption within the depleted layer.

2.2 Excited states

Although the above description in terms of a single differential equation seems to work well in most cases, it is an over-simplification because it neglects the possible existence of excited states. There is still very little understanding about this question, nevertheless, excited states have certainly been observed in few cases. An example is the identification of two-step photothermal transitions in the system Si:S [1]. Another interesting case is the system GaAs:"EL2" [4] (Fig.2), where the excited state is a meta-stable state; the meta-stable property has been explained by very large lattice relaxation [5]. The capture of free carriers in shallow excited states has also been involved in order to explain the experimental observation of capture cross-sections which decrease with increasing temperature [6].

When excited states come into the picture, additional differential equations (one per excited state) must be included in the mathematical analysis.

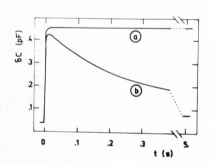

Fig. 2.

Illustration after [4], of the "photo-capacitance self-quenching" effect which occurs in the system GaAs:"EL2" due to the existence of a bound, excited state.
a) T = 200 K, normal photo-capacitance transient; b) T = 80 K, self-quenched photo-capacitance; at the end of the transient all the centers are in an excited, meta-stable state.

2.3 Transitions thermodynamics

We refer to the treatments by Thurmond [7] and by Van Vechten and Thurmond [8].

In the language of thermodynamics, the band gap $\Delta E^O_{cv} = E_c - E_v$ amounts to the increase of Gibb's *free energy* when the number of free electron-hole pairs increases by unity, at constant temperature and pressure:

$$\Delta E^O_{cv} \equiv \left. \frac{\partial G}{\partial n_p} \right|_{T_p} \tag{2.2}$$

It can be demonstrated that ΔE^O_{cv} is the "no-phonon" threshold which can in principle be measured by photo-ionization experiments. ΔE^O_{cv} obeys the thermodynamic law:

$$\Delta E^O_{cv} = \Delta H_{cv} - T \Delta S^V_{cv} \tag{2.3}$$

where ΔH_{cv} is the *transition enthalpy* (or heat of formation at constant pressure) and ΔS^V_{cv} is the *transition entropy*. The following identical equation applies:

$$\frac{d \Delta E^O_{cv}}{dt} \equiv - \Delta S^V_{cv} \tag{2.4}$$

For most semiconductors including Si, Ge, III-V binaries, it is found experimentally that ΔS^V_{cv} is a positive quantity which increases with temperature. Equations (2.3) and (2.4) are illustrated in Fig. 3.

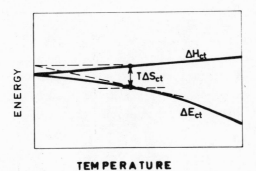

TEMPERATURE

Fig. 3. Illustration of the relationship between free energy ΔE, enthalpy ΔH and entropy ΔS. The extrapolation at $T = 0$ of the tangent line to $\Delta E(T)$ intersects the energy scale at a value equal to the enthalpy $\Delta H(T)$.

In the same way as ΔE^o_{cv}, ΔH_{cv} and ΔS^v_{cv} were defined for the valence - conduction band transition, one can define ΔE^o_{ct}, ΔH_{ct} and ΔS^v_{ct} for, e.g., a bound-to-free transition of an electron from a localized state to the conduction band. Again the thermodynamic equalities

$$\Delta E^o_{ct} = \Delta H_{ct} - T \Delta S^v_{ct} \tag{2.5}$$

$$\frac{d \Delta E^o_{ct}}{dT} = - \Delta S^v_{ct} \tag{2.6}$$

will hold. Again ΔE^o_{ct} can be obtained from optical measurements as the "zero-phonon" threshold of photo-ionization.

The transition entropies ΔS^v_{cv}, ΔS^v_{ct} are of vibrational origin and are related to the softening ($\Delta S > 0$) or stiffening ($\Delta S < 0$) of the lattice at the transition. When an electron is excited out from the valence band and becomes free, there will be a small change in the force constants and correspondingly in the eigen frequencies of lattice vibration. This is reflected in the magnitude and sign of ΔS^v_{cv}. Similarly, for a localized state the electron transition can result in a change of the local phonon frequencies.

In addition to the above discussed "vibrational" entropy, a "degeneracy" of "electronic" entropy also must be considered for the localized states. We recall [9] that, according to Fermi-Dirac statistics, the ratio of occupied states $f_t N_t$ to unoccupied ones $(1-f_t)N_t$ (where f_t is the equilibrium occupation factor and N_t the total number of available localized states) is given by

$$\frac{f_t}{1-f_t} = \frac{g_f}{g_e} \exp(\frac{E_F - E_c + \Delta E_{ct}}{kT}) \tag{2.7}$$

where g_f and g_e are the statistical weights of the filled and empty state respectively. In order to recast (2.7) in the standard thermodynamic form we must redefine the free energy:

$$\frac{f_t}{1-f_t} = \exp(\frac{E_F - E_c + \Delta E_{ct}}{kT}) \tag{2.8}$$

with:

$$\Delta E_{ct} = \Delta E^o_{ct} - kT \ln \frac{g_e}{g_f} \tag{2.9}$$

The term $k \ln \frac{g_e}{g_f}$ adds algebraically to the vibrational entropy ΔS_{ct}^V; it can be either positive or negative. For ordinary (hydrogenic) donors, $g_f/g_e = 2$ and the corresponding contribution to the entropy is *negative*. For hydrogenic acceptors, the contribution is negative too and amounts to $-k \ln 4$.

Using equation (2.8), the equilibrium equality

$$e_n f_t = c_n n (1 - f_t)$$

and the expression of free-electron concentration in a non-degenerate semiconductor

$$n = N_c \exp\left(\frac{E_F - E_c}{kT}\right) \qquad (2.10)$$

one arrives at the well-known "detailed balance" equality

$$e_n = N_c c_n \exp\left(- \frac{\Delta E_{ct}}{kT}\right) \qquad (2.11)$$

and similarly for holes. In a depleted layer, equation (2.11) should be applied with caution, especially with respect to the possibility of field-enhanced emission. From DLS experiments both e_n and c_n can be determined at a given temperature, therefore using eq. (2.11) one can deduce the free-energy ΔE_{ct}. It is also usual to determine the activation energy of thermal emission from an Arrhenius plot of $\lg(e_n/N_c e_n)$ versus $1/T$. Obviously, the quantity determined in this way is ΔH_{ct}, as can be seen by rewriting (2.11) as

$$e_n = N_c c_n \exp\left(\frac{\Delta S_{ct}}{k}\right) \exp\left(- \frac{\Delta H_{ct}}{kT}\right) \qquad (2.11a)$$

with

$$\Delta S_{ct} = \Delta S_{ct}^V + k \ln \frac{g_e}{g_f} \qquad (2.12)$$

Experimentally observed values of ΔS_{ct} for a few deep levels in GaAs are listed in Table 1 [11,12]. It is seen that for all these levels ΔS_{ct} is positive and can be quite large. This situation is also encountered in the other covalent semiconductors, including Si [8]; to our knowledge, negative entropies have not been reported as yet for these materials.

Due to the existence of the electronic, or degeneracy, contribution to the total entropy, the direct comparison of the "optical" free-energy ΔE_{ct}^o with the "thermal" one ΔE_{ct} is affected by an uncertainty, since in general the degeneracy factors g_e and g_f are not known.

Table 1. Experimental enthalpies and entropies for several deep traps in GaAs

Transition	Temp.range K	ΔH eV	$-\Delta S$ 10^{-4} eV/K	Ref.
valence → conduction	270-350	1.5	4.5	7
"EL2"→ conduction	270-300	0.75	1.9...2.5	11,29
"EL3"→ conduction	240-280	0.41	1.9	11
valence → "HL5"	200-220	0.43	4.0	11
valence → "HL2"	310-350	0.71	1.8	11
valence → "HL1"(Cr)	330-370	0.89	>4.5	12

2.4 Lattice relaxation

By lattice relaxation one means the change which occurs in the equilibrium position of the atoms, around a deep state defect or impurity, when the state is filled - as compared to the case when it is empty.

Lattice relaxation is due to the existence of an interaction between the electron and the core motions. Different types of interaction are possible. As a consequence, of the electron-core coupling, the electronic energy E_e is a function of the instantaneous atom positions (in the so-called adiabatic approximation). Assuming a locally dominant atom vibration mode represented by the generalized coordinate ("configuration coordinate") Q, as a first approximation E_e depends linearly on Q:

$$E_e = E_o - bQ \qquad (2.13)$$

where E_o and Q = O correspond to the equilibrium atom position with the electron in the valence or conduction bands.

The new equilibrium position, with the electron in the deep state must correspond to the minimum of the potential energy for atom motion, which in term is equal to the elastic energy plus the energy of the electronic motion:

$$E_c = E_e + \frac{1}{2} KQ^2 = E_o - bQ + \frac{1}{2} KQ^2 \qquad (2.14)$$

By differentiating with respect to Q the equilibrium is at:

$$K\overline{Q} = b$$

The change in *electronic energy* from $Q = 0$ to $Q = \overline{Q}$ is:

$$\Delta E_e \equiv E_o - E_1 = K\overline{Q}^2 \tag{2.15}$$

while the change in *total energy* (electronic + elastic) is only *half* this value. Currently used connotations for this important quantity are:

$$\Delta E_t = \frac{1}{2} K\overline{Q}^2 \equiv d_{FC} \equiv S\hbar\omega \tag{2.16}$$

d_{FC} is called the *Franck-Condon shift*, while S is the *Huang-Rhys factor*; $\hbar\omega$ is the quantum of dominant vibration mode energy. The situation is illustrated in Fig. 4.

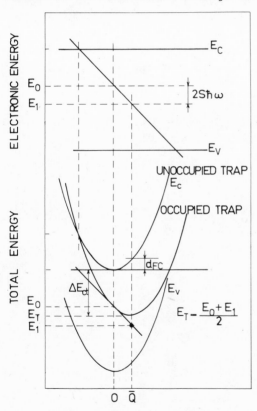

LATTICE DISPLACEMENT (Q)

<u>Fig. 4.</u>   Illustrating the concepts of electron-lattice interaction and lattice relaxation.

Note that in Fig. 4 the force constants for core vibration, in the occupied and in the unoccupied states, have been assumed to be the same (the parabolas have identical shape). Actually, as discussed in the preceding section, the lattice might get softened (or stiffened) at the transition and so more generally one should consider a different force constant, say K_o, in the unoccupied case as compared to K for the occupied one.

As for as Deep Level Spectroscopy is concerned, the main implications of lattice relaxation are:

a) The thermal emission barrier ΔE_{ct} is different (smaller) from the vertical transition (optical emission) energy, by an amount d_{FC} (although the "zero-phonon" photo-ionization threshold is still ΔE_{ct}).

b) When the lattice relaxation phenomenon is pronounced, the probability of optical transitions greatly decreases and is spread out
over a wide range of energies so that the optical bands get very
broad. However, the thermal emission barrier remains of course unique
and equally well defined whatever the strength of the relaxation effect
might be.

c) For very large lattice relaxation, the exchange of carriers
between the free-carrier bands and the localized state can become extremely weak at low temperatures; meta-stable localized states are
thus created. The case has been observed in GaAs as mentioned above,
but also in many other semiconductors.

2.5 The depleted layer

In addition to the physical understanding of the localized electronic states and of the electronic transitions, in order to be able
to analyze correctly the DLS experiments one must also consider
carefully the geometry of the device under test, and make use of such
general relationships as the Poisson equation and the free carrier
drift and diffusion laws of motion.

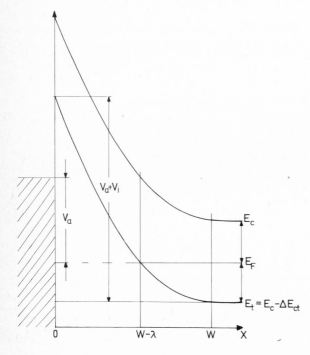

Fig. 5.

Potential variation in
the depleted layer of a
Schottky barrier, and
geometrical definition
of the transition layer
of width λ.

For completeness and further reference, the well-known image of
the potential variation in a Schottky barrier is shown in Fig. 5. We
assume the presence of shallow donor levels (not represented) and of a

single deep donor level. The most useful mathematical relations are:

   a) total potential barrier (internal + reverse bias applied)

$$V_a + V_i = \frac{q}{\varepsilon} \int_0^W [N_D + (1-f)N_T]\, x\, dx \qquad (2.17)$$

   b) total fixed electric charge in the depleted layer

$$Q = \varepsilon\, F(o) = q \int_0^W (N_D + (1-f)N_T)\, dx \qquad (2.18)$$

A very important parameter is the "transition region" of width λ; from Fig. 5 it is clear that λ depends on the energy level depth below E_F, the Fermi level in the bulk. In a steady-state situation, λ is determined by the intersection with the Fermi level as shown, but under dynamic conditions this is no more true. The point will be discussed in more detail below (section 5.1).

3. CLASSIFICATION OF DEPLETED LAYER SPECTROSCOPY METHODS

The basic concepts of Depleted Layer Spectroscopy can be effectively realized in many different ways. Rather than attempting to enumerate and describe them all, it seems more useful and feasible to outline a few criteria of classification as follows:

3.1 According to the type of transition (criterion E):
E1) thermal emission (variable parameter $x \equiv T$);
E2) optical emission (variable parameter $x \equiv h\nu$);
E3) field emission (variable parameter $x \equiv F$).
This criterion has been already discussed. Actually no practical realization of type E3 spectroscopy has been reported yet but field emission is also important as a secondary (perturbing) phenomenon in type E1, and also possibly in type E2, experiments.

3.2 According to the type of device under test (criterion D):
Here we have a two-dimensional matrix:
D1) two-electrode or three-electrode device;
D2) Schottky barrier, p-n junction, or MIS structure.
In all there are six different groups of methods.

3.3 According to the quantity which is being monitored (criterion P):
P1) reverse current;
P2) high-frequency capacitance (i.e., depleted layer with w)

at constant reverse bias; or, by using a feedback loop, reverse bias
at a constant capacitance;

P3) complex admittance $\underline{Y} = G + j\omega C$

3.4 According to the type of excitation waveform (criterion W):

W1) dc or steady-state methods;

W2) small-signal ac methods;

W3) large-signal ac (i.e., pulse/transient) methods.

3.5 According to the "stationarity" feature (criterion S):

S1) non-stationary methods;

S2) stationary methods.

In this context, the stationarity we refer to is relative to the
variation of the spectroscopical variable, i.e. temperature, photon
energy, or field. The stationary methods are characterized by the fact
that the result of the measurement does not depend on the sweeping
rate of the variable (within reasonable limits). The typical example
of a non-stationary method is TSC (Thermally Stimulated Current). On
the other hand, Admittance Spectroscopy, or DLTS (Deep Level Transient
Spectroscopy) are stationary methods. The earlier steady-state photo-
capacitance methods can be qualified as non-stationary, while the more
recently developed ones, which use dual light sources or a combination
of optical and thermal excitation, are examples of stationary methods:
see the paper by H.Grimmeiss at this conference.

The stationary methods are far superior to the non-stationary
ones, with respect to convenience, flexibility, and signal-to-noise
ratio.

4. THE STATIONARY THERMAL EMISSION METHODS

4.1 Deep Level Transient Spectroscopy (DLTS)

Deep Level Transient Spectroscopy is based on the previously
established technique of transient capacitance, but it is a radically new
idea which has been a major contribution to the present revivial of DLS.

A part of the novelty and power of DLTS comes from the electronic
differentiation feature, $C(t_2)-C(t_1)$, used by the inventor D.V.Lang
in his original implementation of the method [14]. However, the main
part of it comes from the *periodic repetition of the excitation*, resulting in a stationary waveform.

Many variations of the idea are possible and have been successfully used in the study of both majority and minority carrier states.

Listed below are a few which have proved most effective:

- electrical excitation with fixed or variable length pulses, allowing the determination of free carrier capture cross-sections [15,16];

- electrical excitation with minority carrier injection, applicable to pn junctions only, allowing the study of minority carrier traps [14,17];

- optical excitation, $h\nu < \Delta E_{cv}$ [18,20] or $h\nu > \Delta E_{cv}$ [19], allowing the study of minority carrier traps with Schottky barriers;

- electronic double differentiation to achieve, in addition to the emission rate window, a well-defined space window [22].

Furthermore it is clear that the principles of DLTS can be extended to *transient current* (instead of transient capacitance) [23,24], and also to three electrode test devices instead of simple diodes [25]. The transient current allows one to greatly extend the upper limit of measurable emission rates. The three-electrode, field-effect transistor-like structures can be driven with extremely fast pulses; a typical example of application would be the determination of large carrier capture cross-sections.

In addition, some of the concepts involved in DLTS have been successfully extended to optical emission characterization [21].

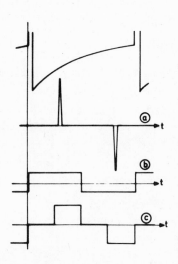

Although the original Lang's method of electronic differentiation and rate window generation is still widely used, more efficient time-filtering techniques, both analog [2,26] and digital [27,30] have been proposed. In the analog filtering class, the use of a lock-in detector [2,28] is an attractive solution, especially with the "P-Q" modification [29] which consists in the substraction of the quadrature (Q) component from the in-phase (P) one at the output of a two-phase, square-wave

Fig. 6. DLTS transient and three different kinds of filtering waveforms; a) double Dirac pulses (Lang); b) square-wave (lock-in detector); c) "P-Q" waveform obtained by subtracting from waveform "b" a similar square wave in quadrature.

lock-in detector. The effective filtering waveform in this case is shown in Fig. 6. As compared to the basic lock-in method, this modification has the following advantages: (i) better resolution, (ii) does discriminate against the excitation pulse, (iii) can be used not only for capacitance transients but also for current transients.

Table 2. Working formulae for DLTS with three types of filtering waveforms. e_o = emission rate, rel.resp. = relative response for a transient of amplitude unity, $\Delta e/e$ = resolution expressed as the full width at half of maximum height.

		"S(t₁)-S(t₂)"		"Lock-in"	"P-Q"
		$t_2/t_1=3$	$t_2/t_1=10$		
transient capacitance	e_o	$1.65/t_2$	$2.56/t_2$	2.51f	1.72f
	rel. resp.	0.39	0.70	0.204	0.152
	$\frac{\Delta e}{e}$	2.70	3.95	3.64	2.65
transient current	e_o	$3.67/t_2$	$10/t_2$	not	3.76f
	rel. resp.	0.27	0.37	appli-	0.403
	$\frac{\Delta e}{e}$	1.90	2.41	cable	1.93

In Table 2, the basic working formulae relative to the emission rate window, the sensitivity, and the resolution are compared for the original Lang method, the lock-in method, and the modified "P-Q" lock-in method.

Although the analog filtering methods tend to be cheaper, investment-wise, and are quite satisfactory, one can expect a more efficient use of information with digital filtering; that is, the output waveform is digitized and fed into a computes which analyzes it in different ways. The fundamental requirement for such a system is very fast data acquisition. A typical application for a fast, computerized set-up consists in calculating the output DLTS signal with *several* different emission rate windows, from the transient measured at a given temperature - instead of repeating the temperature sweep for each window. With this approach, the characterization process of a new semiconductor sample takes less time and can be more accurate.

4.2 Admittance Spectroscopy

Admittance Spectroscopy consists in measuring the a.c. admittance $\underline{Y} = G + j\omega C$ of the depleted layer, as a function of temperature T and angular frequency   .

Historically, the measurement of $C(T,\omega)$ is one of the oldest DLS techniques [31]. The extension to $G(T,\omega)$ is more recent. Ferenczi [32] is one of the pioneers of this last method. A comprehensive up-to-date survey, including original results, is included in the Dr. Sci. thesis by Vincent [33].

Admittance Spectroscopy shares with DLTS the important advantage of stationarity. It yields a spectrum of peaks for $G(T,\omega)$; the capacitance spectrum is made of steps and is well suited for quantitative determinations of the deep state concentrations N_T.

The emission rate window depends slightly on the excitation conditions. For really small-signal excitation, the window is centered at $e = 2\omega$, but in more practical situations it comes nearer to $e = \omega$.

An interesting feature is the existence of a natural "space window": the useful signal is generated in a small fraction Δx of the depleted layer, centered at $x = w - \lambda$, at a fixed distance from the end of the layer (a similar situation is obtained in DDLTS [22], but in this last case the position of the space window with respect to the end of the depleted layer will vary as a function of the bias and pulse voltages). As a consequence, when the bias voltage is varied, the electric field in the space window remains constant as long as N_D does not vary.

Admittance Spectroscopy gives access to the measurement of relatively large emission rates, thus it is particularly useful for shallow traps, and it complements capacitance DLTS in this respect, although transient current DLTS can also do the job. Another specific application is in high-resistivity layers, where it can characterize the Fermi-level-pinning state [34].

Although it possesse some of the desirable properties of DLTS and it is cheaper, the admittance method is far less flexible.

Some of the deeper traps, which can be detected by capacitance DLTS, are not detected in admittance a limitation related to the leakage current.

The carrier capture cross-sections cannot be measured, at least not directly; indirect determinations based on the evaluation of the transition layer width λ are possible in principle, but their dependability has not been proved yet.

No minority carrier traps can be observed.

The widths of both the emission rate window and the space window cannot be varied at will.

5. RECENT DEVELOPMENTS IN THE ANALYSIS OF DLS EXPERIMENTS

5.1 Dynamic behaviour of the transition region

The usual definition for the transition region $w - \lambda < x < w$, relative to a majority carrier level, has been illustrated in Fig. 5. The abscissa $x = w - \lambda$ corresponds to the intersection of the (quasi-) Fermi level with the deep state level. At this abscissa, the capture and emission rates for the majority carriers (electrons in the figure) are equal:

$$e_n = c_n n(x) \qquad at \quad x = w - \lambda \qquad\qquad (5.1)$$

The above geometrical interpretation of λ applies only to majority carrier levels ($e_n \gg e_p$). However, it is possible and useful to define a transition region also in the case of minority carrier levels ($e_p \gg e_n$). While at equilibrium these levels are filled with electrons, with suitable external excitation they can acquire holes. These states will then return to equilibrium with the characteristic rate ($e_p + c_n n$). In close analogy with (5.1), we define the transition region width for minority traps, λ_{minor}, from the equality of the two terms:

$$e_p = c_n \cdot n(x) \qquad at \quad x = w - \lambda_{minor} \qquad\qquad (5.2)$$

Since a long time already, it had been realized that, from the experimentally determined λ of a given majority carrier state, its energy can be obtained [35]. However, the methods used for the determination of λ, based on static capacitance measurements, asked for special samples with very large trap concentrations and extremely good homogeneity; thus their scope was very limited and no conclusive results could be reported.

Meanwhile, transient capacitance studies of carrier capture cross-sections in Si and GaAs had met with difficulties related to long non-exponential tails of the capture transients. Brotherton [36], then - in a penetrating study - Zylbersztejn [37], were able to prove that in many cases these tails were due to the broad distribution of capture rates prevailing in the transition region. A method for the determination of λ, based on dynamic (DLTS) measurements of the trap refilling process, was proposed [37].

The wide variation of the time constants in the transition region is obvious from their explicit expression as a function of abscissa x:

$$\tau^{-1} \equiv r = e_n + c_n(x) = e_n \left\{ 1 + \exp\left[\frac{\lambda^2 - (w-x)^2}{2L_D^2} \right] \right\} \qquad (5.3)$$

where

$$L_D^2 = \frac{kT}{q} \frac{\varepsilon}{qn_o} \qquad (5.4)$$

(n_o = free electron concentration in the bulk) is the Debye length.

We shall now describe a simple and accurate experimental method for the determination of the transition region width, λ, from which the majority carrier capture coefficient, c_n, and the true free-energy depth of the localized state, ΔE_{ct}, can be easily calculated. The method applies in the case of minority carrier levels too, i.e. λ_{minor} and the *majority* carrier capture coefficient c_n can be determined, while of course the energy cannot.

This method, which correctly takes into account the dynamic behaviour of the transition region, has been recently developed by one of the authors [29]. At L.E.P. the method has been applied to several deep states in GaAs [29,12].

The procedure consists of:

a) running a transient capacitance DLTS-type experiment in the way recommended by Lang [14], i.e., with majority carrier pulses of height $\Delta V < V_o$ and duration t_p, superimposed on the reverse bias V_o (Fig. 7). The time interval t_f between two successive pulses is long ($e_n t_f \gg 1$), or else it has to be taken into account in the mathematical analysis;

b) plotting the amplitude of the output signal of transient capacitance, ΔC, against the pulse height ΔV;

c) analyzing the result in the manner outlined below.

The simple analysis which follows is strictly valid only for $N_T \ll N_D$.

Let w_o be the depleted layer width for $V = V_o$, and $w_o - \Delta w$ the width for $V = V_o - \Delta V$. In the present discussion, and due to the condition $N_T \ll N_D$, the tran-

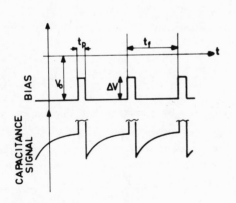

Fig. 7. Definition of the bias waveform for the ΔC vs ΔV experiment.

sient variations of the depleted layer width are considered negligible as compared to the variation Δw. During the pulse, the deep levels situated between $x = w_o - \lambda$ and $x = (w_o - \Delta w) - \lambda_p$ are refilled with electrons. The key feature of the analysis resides in that λ_p is different (smaller) from λ and is a function of the pulse duration t_p.

We calculate λ_p from the condition that the deep states are just half-filled at $x = w - \lambda_p$ at the end of the pulse:

$$\exp(-r.t_p) = \frac{1}{2} \qquad\qquad at \qquad x = w - \lambda_p \qquad\qquad (5.5)$$

Substitution of r from (5.5) and of $w - x = \lambda_p$ in (5.3) yields:

$$\lambda_p^2 = \lambda^2 - z_p^2 \qquad\qquad (5.6)$$

where
$$z_p^2 = 2L_D^2 \ln(\frac{2 - t_p e_n}{t_p e_n}) \qquad\qquad (5.7)$$

Using a stair-case approximation, the output signal can be calculated, see section 2.5, from:

$$\Delta C = const. \int_{W-\lambda_p}^{W_o-\lambda} N_T x dx = const. \left\{ (w_o-\lambda)^2 - (w_o - \Delta w - \lambda_p)^2 \right\} \qquad (5.8)$$

It is seen from (5.8) that for too small a Δw, no signal ΔC is obtained ($\Delta C \geqslant 0$). The threshold value of Δw, called Δw_{thp} is obtained from (5.8) by inspection:

$$w_o - \lambda = w_o - \Delta w_{thp} - \lambda_p$$

or:
$$\Delta w_{thp} + \lambda_p = \lambda \qquad\qquad (5.9)$$

Eliminating λ_p between (5.6) and (5.9), one gets finally:

$$\lambda = (\Delta w_{thp}^2 + z_p^2)/(2\Delta w_{thp}) \qquad\qquad (5.10)$$

Once λ is known, c_n and ΔE_{ct} are calculated:

$$c_n = (e_n/n_o) \exp(\lambda^2/2L_D^2) \qquad\qquad (5.11)$$

$$\Delta E_{ct} = (E_c-E_F) + (E_F-E_T) = kT\ln(N_c/n_o) + (q^2/2\mathcal{E}) n_o \lambda^2 \qquad (5.12)$$

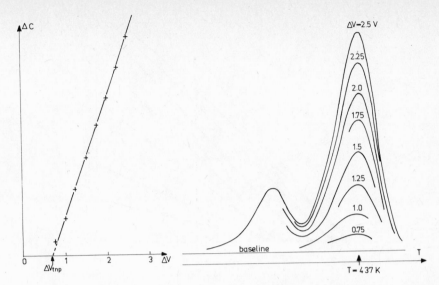

Fig. 8. *Right*: actual DLTS plots with several ΔV pulses for a GaAs:Cr sample. Parameters are $V_o = 5.9$ V, $t_p = 50$ μs, $t_f = 8.2$ ms ("P-Q" filtering) *Left*: plot of the ΔC peak amplitudes against ΔV, illustrating the good linearity and the definition of the threshold ΔV_{thp}.

Experimentally, a threshold ΔV_{thp} is determined by extrapolation from the plot of ΔC versus ΔV, as shown in Fig. 8. Although not obvious by simple inspection from the mathematical relations, it turns out actually that the plot is quite linear over a wide range of ΔV, and the threshold can be obtained with good accuracy. Then, Δw_{thp} is calculated from

$$\Delta w = (2\varepsilon/qn_o)^{1/2} \left[(V_o + V_i)^{1/2} - (V_o + V_i - \Delta V)^{1/2} \right] \tag{5.13}$$

where V_i is the built-in potential.

The above outlined "stair-case" approximation was checked against a rigorous numerical calculation [38]. The agreement is perfect over the full range of possible t_p values, except at the limit $e_n t_p = 1$ where slight difference occur (Fig. 9). Experimentally, by varying the pulse duration t_p one can check the self-consistency of the method and improve the accuracy of λ determination. It is seen from Fig. 10 that the consistency is very satisfactory.

At the time of writing the model has been applied to several deep levels in GaAs, including the native defect "EL2" [29], the electron irradiation-induced defect "E3" [29,38], and the hole level due to c_r [12]. In all cases, it has worked well. The true free-energy level of EL2 was determined in the range 357-400 K (Fig. 11). The obtained values are not inconsistent with the enthalpy of about 0.75 eV obtained

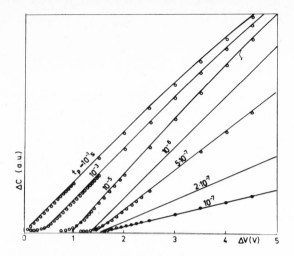

Fig. 9. Comparison between the "stair-case" approximation (full lines)
and a rigorous numerical calculation (dots) of transition
layer refilling kinetics. Parameters are e_n^{-1}=9.53 10^{-2}s,
V_o=5.9 V, $(c_n n_o)^{-1}$=5 10^{-7}s, E_F-E_T=0.196 eV.

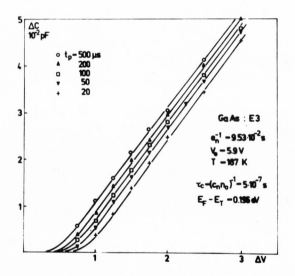

Fig. 10. Comparison of actual experimental data (symbols) with the
numerical calculation (full lines) for different values of
the pulse duration t_p.

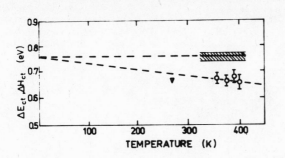

Fig. 11.

Free-energy data for the system GaAs:"EL2" obtained by the method of transition layer dynamics ($\Delta C-\Delta V$ plot). The uncertainty limits are indicated. A previous determination by a different technique [37] is also shown as a triangle. The enthalpy obtained from independent determination is indicated; by calculation one obtains $\Delta S_{et} = 2.5 \ 10^{-4} \ eVK^{-1}$.

from the emission rate signature in the same temperature range coupled with electron capture cross-section measurements extrapolated from lower temperatures [15,20]. The capture coefficient c_n and the related capture cross-section σ_n are in very good agreement (within a factor of two) with the extrapolation of direct determinations (Fig. 12).

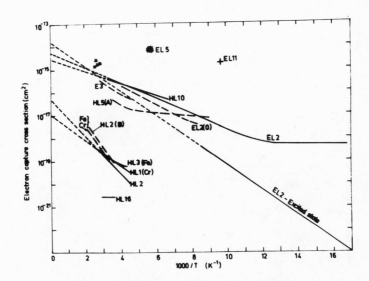

Fig. 12. Electron capture cross-sections for several electron and hole traps in GaAs according to [20] (full lines and symbols) and [15,16] (heavy dotted lines). The indirect determinations for EL2, obtained by the $\Delta C- \Delta V$ plot, are shown as circles.

For the Cr hole level in GaAs, similar data obtained in the temperature range 390-437 K are presented in Fig. 13. The free-energy

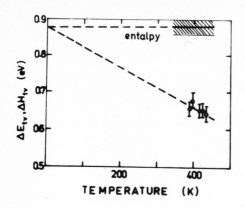

Fig. 13.

Free-energy data for the system GaAs:Cr obtained by the method of transition layer dynamics ($\Delta C - \Delta V$ plot). The enthalpy obtained from independent measurements is also shown. The calculated entropy $\Delta S_{tv} \simeq 5.5 \cdot 10^{-4}$ eVK^{-1}, larger than the one of the bandgap.

level ΔE_{tv} has a strong temperature dependence, as already mentioned in section 2.3. The data is consistent with the enthalpy (0.88 ± 0.02 eV) obtained from the emission rate signature. Again, the calculated hole capture cross-section agrees within a factor of two or so with the direct measurements by Lang [15] and Mitonneau [20] (Fig. 14).

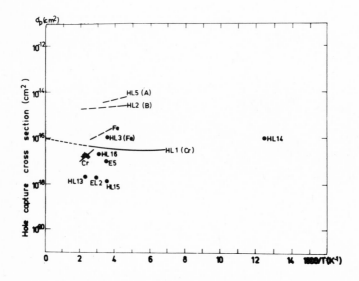

Fig. 14. Hole capture cross-sections for several hole and electron traps in GaAs according to [20] (full lines, full dots) and [15] (heavy dotted lines). The indirect determinations for the Cr trap, obtained by the $\Delta C - \Delta V$ plot, are shown as crosses.

Finally let us now consider the case of minority levels (e.g. a hole level in n-type semiconductor. As already mentioned, in close analogy with the case of majority levels, a transition region exists where the minority carrier (hole) emission process of rate e_p competes with majority carrier (electron) capture; see (5.2). In a hole emission DLTS experiment and ΔC vs. ΔV plot, the effect of λ_{minor} is exactly the same as the one of λ as described above. However, now ΔC is the *signal destruction* plotted against majority-carrier pulse height ΔV. As an example, consider figure   in Lang's original paper [14]: the existence of the ΔV_{thp} threshold is clearly seen. The threshold value can be used exactly as before to obtain c_n and σ_n, only replacing e_n by e_p in (5.7) and (5.11).

5.2 Phonon-assisted tunnel electron emission in GaAs

The enhancement of bound-to-free thermal emission due to the electric field has been recognized since the early deep of Depleted Layer Spectroscopy.

Previously the invoked mechanism has generally been Poole-Frenkel barrier lowering. However, no quantitative agreement with experiment could be reached [39], and it turned out that, at least in some cases, the effect was only an artefact due to the neglection of the influence of the transition layer [40].

In more recent work with GaAs, it has been observed [41] that an electric field - induced enhancement of the emission rate valley exists and is much stronger for electron traps than for hole ones. This directed the efforts towards an explanation based on tunnelling.

Another important observation that had to be explained was a strong temperature dependence of this field effect.

The problem has been recently attacked by two independent and distinct but nevertheless somewhat related approaches. Vincent [33] developed a model based on the concept of virtual states, a kind of thermal equivalent of the Franz-Keldysh effect adapted to deep states, while Makram-Ebeid and Pons worked out a model centered on lattice relaxation [29,42].

These theories are very recent. Both have been partially checked against experiment. It is probable that both would need to be taken into consideration in a unified view which, together with the Poole-Frenkel contribution, would explain every experimental situation.

Pons and Makram-Ebeid have carried out a very detailed comparison with experiment in a well-controlled case. In the following we give a brief account of this theoretical and experimental work.

The lattice relaxation concept has been explained in section 2.4 above. Due to the electron-lattice interaction, the electron level moves slowly up and down in the gap with an amplitude bQ_m (Fig. 4), where Q_m is the amplitude of the lattice vibration:

$$E_{electron} = E_1 - bQ_m \cos \omega t \qquad (5.14)$$

The amplitude Q_m is related to the average number of phonons, n; as follows:

$$\tfrac{1}{2}KQ_m^2 = \hbar\omega \left(n + \tfrac{1}{2}\right) \qquad (5.15)$$

Using (5.15) as well as the definition of the Hwang-Rhys factor S in section 2.4 one can also express bQ_m as follows:

$$bQ_m = 2\hbar\omega \left[S\left(n + \tfrac{1}{2}\right) \right]^{1/2} \qquad (5.16)$$

The slowly varying harmonic term in (5.14) appears as a phase modulation in the "adiabatic" electron wave-function. By developing this function in Fourier-Bessel series, an infinite number of quasi-stationary levels with energies $E_1 \pm p\hbar\omega$, p integer, are obtained; the partial wave-function amplitudes corresponding to each of these levels turn out to be proportional to $\gamma_p(\frac{bQ_m}{\hbar\omega})$, where $\gamma_p(x)$ is the Bessel function of the first kind and order p.

The probability of occupation of each of these quasi-levels depends on temperature, the link being established through (5.16) and the Bose-Einstein statistics of phonon distribution. The following expression for the probability of occupation π_p is obtained:

$$\pi_p = \left[1 + \exp\left(-\tfrac{\hbar\omega}{kT}\right)\right] \sum_{n=o}^{\infty} \exp\left(-\tfrac{n\hbar\omega}{kT}\right) \gamma_p^2 \left(2\left[S\left(n + \tfrac{1}{2}\right)\right]^{1/2}\right) \qquad (5.17)$$

An example of probability distribution π_p is shown in Fig. 15. It is symmetrically centered on E_1.

Knowing π_p one is left with the problem of calculating the tunneling transition rate Γ_p from the pth quasi-level to the conduction band. The electron must tunnel through a potential well of triangular shape, having a height Δ_p where:

$$\Delta_p = E_c - (E_1 + p\hbar\omega)$$

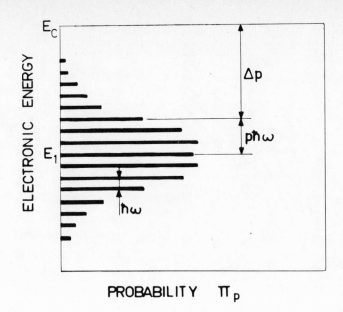

<u>Fig. 15.</u> Example of calculated probability of residence in the
quasi-stationary levels $E_1 + p\hbar\omega$.

and a width $\Delta_p/(qF)$, where F is the local electric field. Recently
the problem was studied by Korol' [43] with the result

$$\Gamma_p = \Gamma \ (\Delta_p, F) = \gamma \frac{\Delta_p}{\hbar K} \exp(-K) \tag{5.18}$$

where

$$K = \frac{4}{3} \frac{(2m^*)^{1/2} \Delta_p^{3/2}}{q\hbar F} \tag{5.19}$$

m^* is the conduction band effective mass, and γ a numerical constant
of order unity which depends on the shape of the potential well.

With π_p and Γ_p defined above, the phonon-assisted tunneling
emission rate e_F at any point in the depleted layer can be written as

$$e_F(x) = \sum_{p=-\infty}^{+\infty} \pi_p \cdot \Gamma \left[\Delta_p, F(x) \right] \tag{5.20}$$

This model has been thoroughly tested in the case of the irradiation-induced defect "E3" in GaAs [44]. The numerical evaluation of
(5.20), with parameters which are adequate for this particular deep
state, is illustrated in Fig. 16. It can be appreciated that the effect

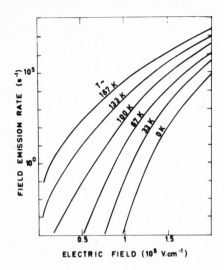

can be quite strong already at not too high fields ($< 10^5$ Vcm^{-1}). An approximate analytical treatment shows that the effect depends mostly on the product ST, so that for large lattice relaxation it is already strong at low temperatures and vice-versa.

In order to test the theory against experiment, the emission rate (5.20) must be integrated over the depleted layer. A wide range of time constants is obtained at any temperature. In Fig. 17, the experimental capacitance transients are composed with the calculations at a fixed bias and

<u>Fig. 16.</u> Calculated field emission rate as a function of the electric field and temperature. Parameters are E_1 = 0.35 eV, $\hbar\omega$ = 10 meV, S = 7.5, m^*= 0.068 m_e (n–GaAs)

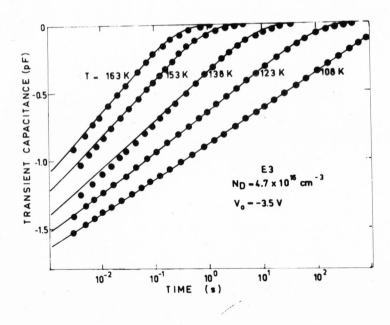

<u>Fig. 17.</u> Calculated (full lines) and experimental (dots) capacitance transients in a Schottky barrier on n–GaAs with electron-irradiation defect E3. The curve for T = 123 K was fitted by varying S and γ.

different temperatures. Good agreement has likewise been obtained as
a function of reverse bias, and also in samples having various doping
levels, as shown in Fig. 18. This figure clearly illustrates the fact
that already at the 10^{16} cm^{-3} doping level the field effect can modify
the deep level "signature" (T^2/e_n vs 1/T) appreciably.

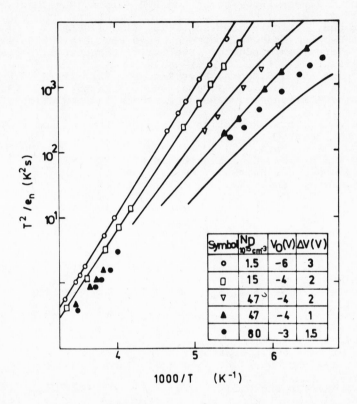

Fig. 18. Experimental (symbols) and calculated (full lines) signatures
          (T^2/e_n vs 1000/T) of the E3 trap in samples with various doping levels
          and with different bias pulses.

The best fit values of the parameters E_1, $\hbar\omega$, and S are entirely
consistent with previous studies of the deep state "E3", in particular
with the electron capture cross-section measurements by Henry and Lang
together with their interpretation of the data in the frame of a
multi-phonon-emission (MPE) model, where lattice relaxation plays
an essential role [16].

REFERENCES

1. C.T.Sah, Solid State Electron. $\underline{19}$, 975 (1976)
2. G.L.Miller, D.V.Lang and L.C.Kimerling, Ann.Rev.Mater.Sci. 1977, 377
3. P.Blood and J.W.Orton, Rep.Progr.Phys. $\underline{41}$, 157 (1978)
4. A.Mitonneau and A.Mircea, Solid State Commun. $\underline{30}$, 157 (1979)
5. D.Bois and G.Vincent, J.Phys.(F) $\underline{38}$, L351 (1977)
6. R.M.Gibb, G.J.Rees, B.W.Thomas and L.H.Wilson, Phyl.Mag. $\underline{36(4)}$, 1021-1034 (1977)
7. C.D.Thurmond, J.Electrochem.Soc. $\underline{122}$, 1133 (1975)
8. J.A. van Vechten and C.D.Thurmond, Phys.Rev. $\underline{B14}$, 3539 (1976)
9. J.S.Blakemore, Semiconductor Statistics, chapter 3, Pergamon Press, 1962.
10. C.M.Penchina and J.S.Moore, Phys.Rev. $\underline{B9}$, 5217 (1974)
11. A.Mircea, A.Mitonneau and J.Vannimenus, J.Phys.(F) $\underline{38}$, L41 (1977)
12. G.M.Martin, A.Mitonneau, D.Pons, A.Mircea and D.W.Woodard, to be published
13. A.Mircea and D.Bois, Radiation Effects in Semiconductors (Nice) 1978 (Inst. Phys. Conf. Ser. 46)
14. D.V.Lang, J.Appl.Phys. $\underline{45}$, 3023 (1974)
15. D.V.Lang and R.A.Logan, J.Electron.Mat. $\underline{4}$, 1053 (1975)
16. C.H.Henry and D.V.Lang, Phys.Rev. $\underline{B15}$, 989 (1977)
17. D.V.Lang, J.Appl.Phys. $\underline{45}$, 3014 (1977)
18. A.Mitonneau, G.M.Martin and A.Mircea, GaAs and related compounds (Edinburgh) 1976 (Inst. Phys. Conf. Ser. 33a)
19. R.Brunwin, B.Hamilton, P.Jordan and A.R.Peaker, to be published
20. A.Mitonneau, A.Mircea, G.M.Martin and D.Pons, Revue de Physique Appliqueé to be published
21. A.Chantre, Thesis Dr és Sci. Lyon 1979
22. H.Lefevre and M.Schulz, Appl.Phys. $\underline{12}$, 45 (1977); also IEEE Trans. Electron Dev. $\underline{ED-24}$, 973 (1977)
23. B.W.Wessels, J.Appl.Phys. $\underline{47}$, 1131 (1976)
24. G.M.Martin and D.Bois, Characterization Techniques for Semiconductor Materials and Devices (Seattle) 1978, vol. 78-3, 32.
25. M.G.Adlerstein, Electron.Lett. $\underline{12}$, 297 (1976)
26. A.Mircea, private communication
27. G.B.Stringfellow and E.E.Wagner, to be published
28. A.Mircea, A.Mitonneau, J.Hallais and M.Jaros, Phys.Rev. $\underline{B16}$, 3665 (1977)
29. D.Pons, Thesis Dr Ing Paris, 1979
30. A.M.White, A.J.Grant and B.Day, Electron.Lett. $\underline{14}$, 411 (1978)
31. C.T.Sah and V.G.K.Reddi, IEEE Trans.Electron Dev. $\underline{ED11}$, 345 (1964)
32. G.Ferenczi, J.Kiss and M.Somogyi, Physics of Semiconductors p.783, 1976 (Rome)
33. G.Vincent, Thesis Dr és Sci Lyon (1978)
34. H.C.Casey, Jr., A.Y.Cho, D.V.Lang, E.H.Nicollian and P.W.Foy, to be published in J.A.P.
35. G.H.Glover, IEEE Trans. Electron Dev. $\underline{ED-19}$, 138 (1972)
36. S.D.Brotherton, private communication
37. A.Zylbersztejn, Appl.Phys.Lett. $\underline{33}$, 200 (1978)
38. D.Pons, to be published
39. J.W.Walker and C.T.Sah, Phys.Rev. $\underline{B8}$, 5597 (1973)
40. H.G.Grimmeiss, private communication
41. D.V.Lang, L.C.Kimerling and S.Y.Leung, J.Appl.Phys. $\underline{47}$, 3587 (1976)
42. D.Pons and S.Makram-Ebeid, to be published
43. E.N.Korol', Sov.Phys. Solid State $\underline{19}$, 1327 (1977)
44. L.C.Kimerling and D.V.Lang, Lattice Defects in Semiconductors 1974 (Inst. Phys. Conf. Ser. 23) p.589 (1975)

LUMINESCENCE OF CHROMIUM IN GALLIUM ARSENIDE*

Claude M. Penchina**
Department of Physics and Astronomy
University of Massachusetts
Amherst Mass. 01003 USA

Research done in collaboration with
Edward C. Lightowlers and Martin O. Henry, King's College, London
Milena Zavetova and Bedrich Velicky, Inst. of Physics, Academy of Sciences, Prague

INTRODUCTION

The luminescence of chromium-doped gallium arsenide was found many years ago (Allen 1968) to exhibit a broad bright band in the near infrared. However, it was not until 1976 that a 0.84 eV zero-phonon line associated with this band was first reported (Stocker & Schmidt 1976, Koschel, Bishop & McCombe 1976). This zero-phonon "line" was soon observed to have fine structure (Lightowlers & Penchina 1978) which has since been resolved into a multiplet of at least 13 zero-phonon lines; 4 of these are easily visible even on a broad scan of this luminescence band shown in Fig. 1. Presumably the fine structure was not observed long ago because of either poor signal-to-noise ratio or inappropriate choice of spectrometer resolution.

An investigation of Cr-doped semi-insulating GaAs from various sources has revealed the presence of a number of other luminescence bands, the most common being a broad band around 2 μm which is generally present in materials doped with oxygen (Fig. 2). In crystals where this oxygen related band is weak or absent, a further sharp line system was observed (Lightowlers & Penchina 1978) around 0.57 eV (2.16 μm) as shown in Fig. 3. This luminescence band was first reported by Koschel et al. (1976) who failed to observe the fine structure of the zero-phonon multiplet. We have observed it also in the LPE sample used by Stocker & Schmidt (1976) and in a variety of bulk samples. We presume this band accompanies the 0.84 eV band in all samples, but may be hard to observe due to problems of signal-to-noise when the oxygen related band is strong.

* Supported in part by the ONR under contract N 00014-76-C-0890, by the Science Research Council, and by the University of London Central Research Fund.

** NAS Exchange scientist, Inst. of Physics, Academy of Sciences, Prague, and Inst. for Technical Physics, Academy of Sciences, Budapest.

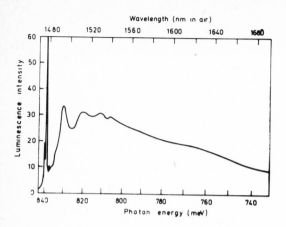

Fig. 1.

Cathodoluminescence spectrum of the 0.80 eV system, un-corrected for the transfer function of the optical system, obtained from Sumitomo SI GaAs:Cr with a cold finger temperature of 4.8 K and a beam current of 2.0 μA. The temperature of the emitting region was 6.0±0.5 K. The no-phonon structure is distorted by the slitwidth of the mono-chromator and system re-sponse time. The main peak at 839.37 meV has a height of 150 on this scale.

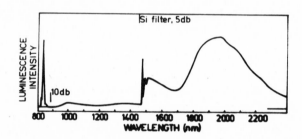

Fig. 2. Cathodoluminescence spectrum of GaAs:Cr produced by Crystal Specialties obtained at ∼10 K, un-corrected for the transfer function of the op-tical system. The long wavelength band is thought to be related to the presence of oxygen.

Fig. 3.

Cathodoluminescence spectra of the no-phonon structure at 0.575 eV obtained from Sumitomo SI GaAs:Cr at (a) 7.7 K and (b) 21.2 K. The energy level scheme in-ferred from the temperature dependence of the relative line strengths is shown as an insert.

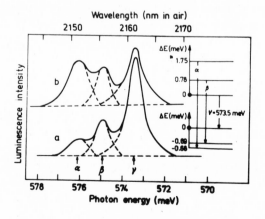

In this lecture, we shall concentrate mainly on the rather rich details of the luminescence between 2.1 and 2.4 μm associated with the zero-phonon triplet near 0.57 eV, and shown in Fig. 4.

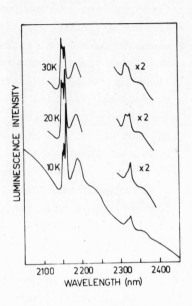

Fig. 4a.

Cathodoluminescence spectra of the 0.57 eV system at ∼10 K, 20 K and 30 K obtained from Sumitomo SI GaAs:Cr. The sharp structure at lower energy appears to be a phonon replica of the no-phonon lines at 0.575 eV (2.16 μm).

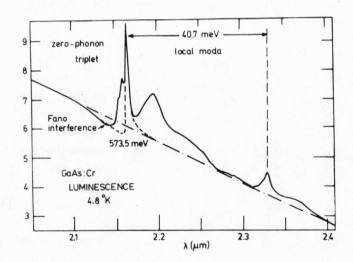

Fig. 4b. Fit of Fano theory (----) to experimental luminescence spectrum (———) where the background is interpolated (-.-.-.) between smooth regions of the spectrum. q= -4, Γ= 0.88 eV, p² = 0.37.

EXPERIMENTAL ARRANGEMENT

Luminescence was excited by an electron beam of 50 keV and a typical beam current of 3 μA. The beam was deflected on and off the sample electronically to allow lock-in detection, so the average power was about 75 mW. The arrangement of the experiment is shown schematically in Fig. 5. The cathodoluminescence seems to be more effective than photo-luminescence excited by visible lasers; the visible light is absorbed so close to the surface that it excites a region of lower quality than the deeper bulk, and also causes some local heating. The luminescence is dispersed through a grating spectrometer and detected with a cooled lead-sulphide photoconductor. An important feature in the study of this spectrum is the optimization of signal-to-noise. This requires that the spectrometer slits be made as wide as possible without degrading the resolution required for the experiment. Since the spectral lines in this region are already broad, this allowed us to use 1mm slitwidth, permitting the detection of features which had been previously unobserved.

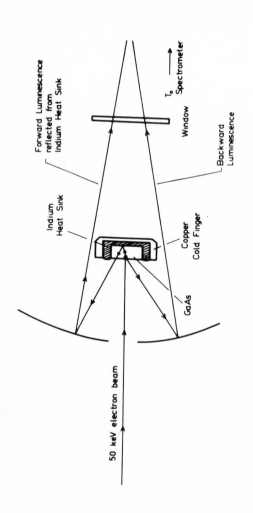

Fig. 5.   Experimental arrangement for cathodoluminescence

RESULTS

Figure 4a shows much interesting structure, of which we shall concentrate on these five features:

A) A triplet of zero-phonon lines around 575 meV.

B) A large broad background signal which decreases towards longer wavelength.

C) A broad system of sidebands to the long wavelength side of the zero-phonon lines.

D) A sharp sideband shifted 40.7 meV to longer wavelength from the zero-phonon lines.

E) A dip in the luminescence intensity on the short wavelength side of the zero-phonon lines.

Feature A, the zero-phonon lines, has been studied to higher resolution (Lightowlers & Penchina 1978) as a function of temperature, and the splittings in the ground state and excited state deduced as shown in Fig. 3. It is generally thought to be due to an excitation of Cr^{3+} (ESR notation for Cr in which 3 electrons have contributed to the valence bonding in the crystal, i.e. neutral chromium when substituting for gallium).

Feature B, the broad background, is due largely to the broad 2 μm oxygen related luminescence band reported by Lightowlers et al. in 1978 (Lightowlers, Henry & Penchina 1978).

Feature C, the broad sidebands, looks like a familiar crystal phonon replica of the zero-phonon luminescence. It seems to be missing a contribution from the large peak in the phonon density of states due to longitudinal optical phonons (Johnson 1966).

Feature D, is a sharp replica of the zero-phonon lines, which was first reported to be a local phonon replica in 1978 (Lightowlers, Henry & Penchina 1978).

Feature E, which appears to be some sort of anti-resonance, has not been previously explained.

In the remainder of this lecture, we shall concentrate on an explanation for features E and B, and for features C and D which will be treated together.

Feature E: anti-resonance

The broad spectral scan (Fig. 4a) of the luminescence shows quite clearly a rather sharp dip in the luminescence intensity, just to the short wavelength side of the zero-phonon lines near 575 meV. One possible explanation which first came to mind was that it might be due

to self absorption of the oxygen related background luminescence (Fig.2)
by a Stokes shifted phonon replica of the zero-phonon Cr lines. Because
of the experimental arrangement used (Fig. 5) in which the luminescence
is observed in the backward direction, this absorption would have to
occur in only a few microns (the electron penetration depth) for the
primary luminescence in the back direction, or in a total of twice the
sample thickness (about 0.4 mm thick) if it were absorption of the
forward luminescence reflected back by the indium heat sink. In either
case, this strong absorption should be easily detected in an infrared
absorption measurement. No such sharp absorption band was observed,
thus ruling out this tentative explanation.

Another explanation is suggested by the relatively large width of
the 0.57 eV zero-phonon lines (FWHM about 1 meV) compared with the
width of the 0.84 eV zero-phonon lines (FWHM about 0.2 meV) (Lightowlers
& Penchina 1978) at liquid helium temperature. This additional width
is in spite of an apparently weaker phonon sideband spectrum. We suggest here that the broadening and the anti-resonant dip are both due
to a degeneracy in energy of a discrete state of the impurity and a
continuum state, a so-called Fano-resonance. Fano has shown (Fano &
Cooper 1968, Velicky & Sak 1966) that when a "discrete" state is degenerate with a continuum, the interaction broadens the discrete level
into a resonant level, with interference terms causing a nearly anti-
-resonance and asymmetric lineshape. The theory predicts

$$\sigma(E) = \sigma_{cont.}(E) \left[\frac{p^2(q+\epsilon)^2}{1+\epsilon^2} + (1-p^2) \right]$$

where

$$\epsilon = \frac{E-E_{resonance}}{\frac{1}{2}\Gamma}$$

Fitting this theoretical expression to our experimental spectra (Fig 4b)
yields a linewidth $\Gamma = 0.88$ meV, which is much larger than kT (about
0.3 meV at 4.2°K), a resonance lineshape given by q = -4, and a coherence with the background given by $p^2 = 0.37$. We fit the theory only
to the strongest zero-phonon line. A fit to all three lines would
require some additional knowledge of their coherence with each other.
The relatively nice fit, as well as the lack of other plausible explanation for the line broadening and anti-resonance, leads us to conclude that there is indeed a Fano interference between a discrete level
and a continuum.

Feature B: oxygen related background

This background luminescence band is illustrated in Fig. 2 for a sample of chromium doped GaAs from Crystal Specialties. The same band is much brighter in their chromium/oxygen doped material, and is the only important feature in their GaAs:O. This band appears also in other samples of GaAs with intentional oxygen doping from RSRE and Sumitomo, and does not appear in an undoped GaAs sample from NRL grown in boron nitride to specifically exclude oxygen contamination (Swiggard et al. 1977). On the other hand, we have studied two samples with intentional oxygen doping which do not show this feature either: one from RSRE shows several other broad bands, while one from Sumitomo shows only luminescence characteristic of Cr, though its photoconductivity shows evidence of oxygen (Tyler, Jaros & Penchina 1977). Thus, we conclude that the broad 2 μm band is evidence of oxygen impurities. The absence of this band does not, however, necessarily prove the absence of oxygen, which might enter GaAs in some other state, complex, etc.

Features C and D: low wavelength sidebands

The intensity of the broad long-wavelength sidebands of the 0.57 eV zero-phonon lines (Fig. 6a) looks quite similar to the density of lattice phonons of GaAs (Johnson 1966) (Fig. 6b). A major difference is that the peak in the density of states due to longitudinal optical phonons does not appear in the sideband spectrum. There is instead a peak shifted by 40.7 meV to longer wavelength which has the same shape as the zero-phonon triplet, and thermalizes with it. If one assumes the peak in the density of phonon states can be represented by a single frequency (i.e. Einstein spectrum), then the shifted peak is explained quite well as a local vibration of Cr on a Ga site with no major change in force constants. Since $W = (k/m)^{\frac{1}{2}}$, the ratio of the frequencies would be

$$\sqrt{\frac{M_{Ga}}{M_{Cr}}} = \sqrt{\frac{69.72}{52.00}} = 1.16$$

which is in quite good agreement with the measured ratio $\overline{W}_{local}/\overline{W}_{peak} = 1.18$.

A more accurate treatment (Penchina et al. 1979), using the theory of Dawber and Elliott (1963) extended to the case of a compound semiconductor, determines the local mode frequency from the mass defect and an integral over the full lattice phonon spectrum. This, however, gives a local mode frequency which is too low to give a good fit to the experiment (Fig. 6c). A sufficiently high frequency could be obtained

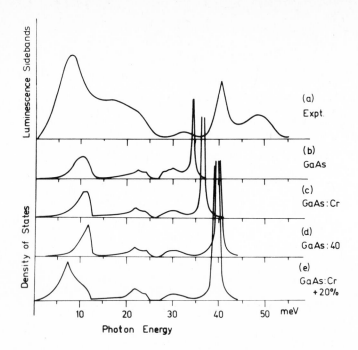

Fig. 6. a) Experimental phonon sidebands of 0.57 eV GaAs:Cr luminescence after
         removal of zero-phonon lines and background luminescence. The structure
         beyond the 40.7 meV peak is due to emission of a local phonon and a
         band phonon.
      b) Lattice phonon density of states of GaAs, from Johnson (1966)
      c) Density of phonons at defect for Cr on Ga site, no change of force
         constants.
      d) Density of phonons at defect for mass 40 on Ga site, no change of force
         constants.
      e) Density of phonons at defect for Cr on Ga site, 20 % increase in local
         force constants.
      The vertical scales of 6(a)-(e) are arbitrary, and vary from curve to
      curve.

by using a mass of about 40 atomic units (a rather unlikely impurity
mass), at the cost of getting a worse fit to the low energy acoustic
phonon sidebands (Fig. 6d). On the other hand, for a Cr impurity, a
model using nearest neighbour interactions enhanced by about 20 %
(Penchina et al. 1979) gets the correct local mode frequency and
simultaneously improves the fit for the low energy acoustic phonon
sidebands as well (Fig. 6e).

It might seem at first surprising that an increased force constant
would increase the local mode frequency but decrease the acoustic
phonon frequencies. This is easily explained when one realizes that in

the low energy acoustic modes, nearest neighbours move in phase. Increasing the force constant tends to bind the defect more rigidly to its neighbours, thus causing it to "drag" the neigbours along an effect similar to increasing the local mass, which then lowers the frequency. The increase in force constant could come from the interactions involving the unfilled d shell of Cr, and/or from a difference in charge state between the Cr impurity and the Ga for which it substitutes.

DISCUSSION

Our study of the 0.57 eV luminescence band in GaAs:Cr indicates that the main features can be explained in terms of a triplet of zero-phonon lines of chromium which are degenerate with a continuum, and thus exhibit both broadening, and a Fano type anti-resonance. The optical transitions are coupled to local vibrations of the chromium impurity which exhibits a 20 % increase in local force constants. By comparison, the 0.84 eV luminescence shows no evidence of Fano resonance or coupling to local phonons. Since the coupling to phonons is so different for the 0.57 eV and the 0.84 eV luminescence, they are presumed to be due to two different charge states of Cr. The 0.84 eV luminescence is thought to be in some ways related to Cr^{2+} (i.e. singly negatively charged chromium on a gallium site), perhaps paired with some shallow impurity. Thus, we expect that the 0.57 eV luminescence is related to Cr^{3+} (neutral chromium on a gallium site) which should also be present in semi-insulating gallium arsenide.

The broad luminescence band around 2 μm was found to be characteristic of oxygen in GaAs. There were, however, samples which supposedly contained oxygen which did not show this band. Thus, it is likely that oxygen enters GaAs in more than one state or complex, only one of which produces this luminescence band. Additional study of oxygen-doped and oxygen-free samples will be needed before this luminescence band can be used as a definitive test for oxygen impurities.

REFERENCES

G.A.Allen, J.Phys.D. 1, 593 (1968)
P.G.Dawber and R.J.Elliott, Proc.Roy Soc. 273, 222-236 (1963)
F.A.Johnson, Progress in Semiconductors 9, 181-235 (1966)
W.H.Koschel, S.G.Bishop and B.D.McCombe, Solid State Commun. 19, 521 (1976)
E.C.Lightowlers, M.O.Henry and C.M.Penchina, Int. Conf. on Recombination in Semiconductors, Southampton, U.K. (unpublished) (1978)
E.C.Lightowlers, M.O.Henry and C.M.Penchina, Int.Conf. on Physics of Semiconductors, Edinburgh, Scotland 1978. Inst of Physics Conf. Series 43, 307-310 (1979)

E.C.Lightowlers and C.M.Penchina, J.Phys.C. $\underline{11}$, L405 (1978)

C.M.Penchina, E.C.Lightowlers, M.O.Henry, M.Zavetova and B.Velicky, Proc. RECON conf., Prague (1979) (to be published)

H.J.Stocker and M.Schmidt, J.Appl. Phys. $\underline{47}$, 2450-1 (1976)

E.M.Swiggard, S.H.Lee and F.H.Von Batchelder, Inst. Phys. Conf. Ser. $\underline{33B}$, 23 (1977)

E.H.Tyler, M.Jaros and C.M.Penchina, Appl. Phys. Letters $\underline{31}$, 208 (1977)

B.Velicky and J.Sak, Phys. Stat. Sol. $\underline{16}$, 147 (1966)

ANALYSIS OF DEFECT STATES BY TRANSIENT CAPACITANCE
METHODS IN PROTON BOMBARDED GALLIUM ARSENIDE AT 300 K AND 77 K

A. Nouailhat, G. Guillot, G. Vincent, M. Baldy
Laboratoire de Physique de la Matière*
Institut National des Sciences Appliquées de Lyon
20, Avenue Albert Einstein 69621 VILLEURBANNE Cédex (France)

and

A. Chantre
C.N.E.T.
B.P. n° 42 48240 MEYLAN (France)

ABSTRACT

Thermal capacitive spectroscopy (DLTS, conductance) measurements have been performed on N-type GaAs Schottky barriers irradiated at 300 K and 77 K with one hundred keV protons at dose of some 10^{11} p$^+$/cm^2. At 300 K, the electron traps E_2, E_3, E_4, E_5, found also after electron irradiation, are created together with another trap (E_c-0.3 eV, $\sigma_{na} = 3.10^{-14}$ cm^2). At 77 K, the E_2 and E_3 defects are also created together with a new electron trap (E_c-0.26 eV, $\sigma_{na} = 9.10^{-13}$ cm^2) which anneals between 200 K and 300 K. We have also applied the new method DLOS (Deep Level Optical Spectroscopy) to the determination of the optical capture cross-section $\sigma_n^o(h\nu)$ of E_3 to determine the lattice relaxation effect of this centre.

INTRODUCTION

GaAs is now widely used in modern electronics (laser diode, FET...) and several works have been performed on the deep levels found in this material since the introduction of the DLTS apparatus which allows a real thermal spectroscopy of defects [1,2,3]. In spite of this fact, reliable information concerning the nature and the microstructure of the defects is very limited in GaAs for which ESR spectra give poor results [4].

The irradiation appears as a way of studying relatively simple defects with a number of parameters which can be easily changed: irradiation temperature, energy, dose. Only the defects which are stable at room temperature have been studied. Five electron traps or strictly speaking the levels due to this traps (E_1 to E_5) and one hole trap (H_1)

* Equipe de Recherche associée au C.N.R.S.

have been detected after electron irradiation of liquid phase epitaxy
layers (LPE) GaAs 5 and a great deal of information has been obtained
on their annealing behaviour, the variations of their energy in the
gap as a function of x in compounds like $GaAl_xAs_{1-x}$ and the lattice
coupling [6,7,8]. Recently, Pons has analysed the defects created at
300 K in vapor phase epitaxy (VPE) GaAs by electron irradiation, their
annealing properties and their introduction rate dependence versus
electron energy [9,10,11].

However, for the moment, all the spectroscopic data for the irradiation defects are thermal: we know only the thermal ionization energy
with respect to the conduction or valence band (respectively E_n or E_p)
and the capture cross sections ($\sigma_n(T)$ and $\sigma_p(T)$). All the corresponding
optical data ($E^o n$, E^o_p, $\sigma^o_n(h\nu)$, $\sigma^o_p(h\nu)$) for these levels are unknown,
although these quantities contain very important information about the
physics of the defects (relaxation process, relation to the bands).

Another important problem about the defects created at 300 K is
the following: are they primary defects which would appear after irradiation at low temperature before the annealing stages I and II (at
235 K and 280 K) found by Thommen [12] or are they secondary ones?

The aims of this paper are:

i) to report about the electron traps observed in proton irradiated GaAs (VPE) at 300 K and for the first time at 77 K to see if the
defects created at the two temperatures are the same and if during the
annealing stages I and II the concentration of defects undergoes some
changes, and

ii) to give some optical information on the E_3 trap by using a new
optical spectroscopic method which has been recently set up in our
laboratory by Chantre and Bois [13,14]: the Deep Level Optical Spectroscopy (DLOS).

We have chosen proton irradiation of low energy (~ 100 keV) because
a small proton accelerator is quite well adapted to the "in situ"
characterization of defects at low temperatures. The defects are
created in the depletion region of Schottky diodes and profile measurements are possible by capacitance technics.

EXPERIMENT

Two n-types layers of GaAs grown by VPE ($AsCl_3$ process) have been
used: the sulfur doping levels of the epilayers are $2x10^{16}/cm^3$ and
$5x10^{16}/cm^3$. Before irradiation a 300 Å thick layer of gold is evaporated
on the sample surface which is tilted by 7° with respect to the proton

beam. Proton implantation was performed at room and liquid nitrogen temperatures at an energy of 90 keV with fluxes of $2.5 \cdot 10^{10}$ p$^+$/cm^2.sec and 5.10^{10} p$^+$/cm^2.sec. Taking into account the proton energy loss in the gold layer, the distribution of defects can be characterized by a projected range $R_p = 0.7$ µm [15,16].

Immediately after irradiation, the Schottky diodes are analysed by an experimental apparatus of capacitance transient spectroscopy which can give:

i) the free carrier profile n(x),

ii) the DLTS spectra with window rates e_n ranging from $10^{-1}s^{-1}$ to 300 s^{-1}. This range can be extended up to $10^6 x^{-1}$ either by current transient or conductance measurements, and

iii) the DLOS spectra giving the optical capture cross sections ($\sigma_n^o(h\nu)$ or $\sigma_p^o(h\nu)$) of a deep level by measurements of the initial derivative of photocapacitance transients of a Schottky diode. The initial conditions are fixed: all the levels full or empty at the beginning of the illumination.

EXPERIMENTAL RESULTS

A) Optical capture cross section $\sigma_n^o(h\nu)$ of the E_3 defect

We have measured the DLOS spectra on an electron-irradiated sample at 300 K at two temperatures (110 K and 190 K) that is before and after the DLTS peak of E_3 in the same time constant conditions of measurements. At 110 K, the DLOS spectrum represents the addition of the contribution of all the defects which are non thermally ionized at this temperature: the electron traps (E_3, E_4, E_5) and the hole traps [9]. At T = 190 K, the defect E_3 contributes no more to the DLOS spectrum. The $\sigma_n^o(h\nu)$ spectrum obtained by difference of the two curves at T = 110 K and T = 190 K is shown in Fig. 1.

B) Deep levels created by proton irradiation at 300 K

Figure 2 shows a DLTS spectrum of a proton-irradiated sample at 300 K. It gives evidence of the presence of 5 defects. The signatures of these defects ($\ell_n \frac{e_n}{T^2} : \frac{1}{T}$) are given in Fig. 3. Three of these electron traps are E_3, E_4, E_5 observed by Lang [5] and Pons [9]. The E_2 trap, more shallow, is analyzed by conductance measurements (Fig. 2). Table 1 gives the apparent thermal activation energy and the apparent capture cross section of the five defects.

The new defect D_1 ($E_c-0.3$ eV, $\sigma_{na} = 3 \cdot 10^{-14}$ cm^2) seems related specifically to proton irradiation.

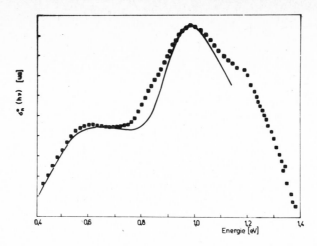

Fig. 1. Optical capture cross section associated to the E_3 level from [13]. ■: spectrum obtained by the difference between DLOS spectra at 110 K and 190 K. ——; theoretical fit with: $E_n^o = 0.44$ eV; $d_{FC} \simeq 0.2$ eV; $\propto^{-1} = 4.5$ Å; $P_L/P_\Gamma = 0$; $P_L^n/P_\Gamma = 0.06$

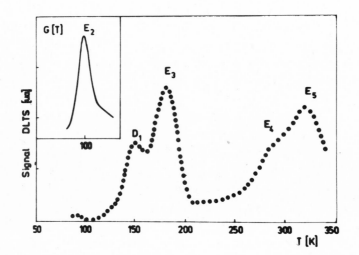

Fig. 2. DLTS spectrum of a GaAs VPE layer irradiated with protons at 300 K (90 keV, 2.10^{11} p^+/cm^2). Window rate: $e_n = 12.8$ s^{-1}. Insert: conductance peak due to the E_2 defect (frequency: 10 kHz)

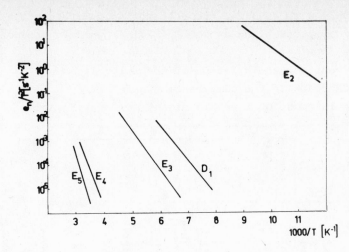

Fig. 3. "Signatures" of the defects created by proton irradiation at 300 K in GaAs VPE (90 keV, 2.10^{11} p^+/cm^2)

Table I. Apparent activation energy and apparent capture cross section for the electron traps created by proton irradiation at 300 K

	E_2	D_1	E_3	E_4	E_5
E_{na} (eV)	0.16	0.3	0.33	0.65	0.85
σ_{na} (cm^2)	7.10^{-12}	3.10^{-14}	3.10^{-15}	$6.8.10^{-14}$	8.10^{-12}

C) Deep levels created by proton irradiation at 77 K

Up to day no spectroscopic experiments have been performed on low temperature irradiation defect in GaAs.

The electron traps created at 77 K have been analyzed by DLTS spectra between 77 and 200 K in order to avoid the annealing stages I and II.

Table II. Apparent activation energy and apparent capture cross section for the electron traps created by proton irradiation at 77 K

	E_2	I	E_3
E_{na} (eV)	0.16	0.26	0.35
σ_{na} (cm^2)	7.10^{-12}	9.10^{-13}	$1.5.10^{-14}$

Figure 4 shows a typical DLTS spectrum measured in these conditions. In the "energy window" explored, it was possible to know the signatures of three created defects, given in Table II and in Fig. 5. Two of these traps are E_2 and E_3 formed by room temperature irradiation. The third level which has been called I is reported for the first time and appears to be present only with liquid nitrogen temperature irradiation.

By thermally annealing at 300 K (Fig. 4) we have produced three main effects:

i) the complete annihilation of the I level,

ii) a decrease in the compensation and

iii) an increase of the DLTS signal $\Delta C/C$ for E_2 and E_3.

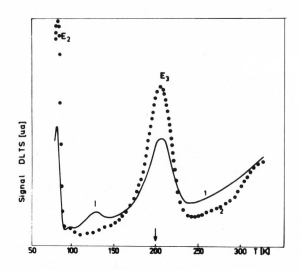

Fig. 4. DLTS spectra of a GaAs VPE layer proton irradiated at 77 K (95 keV, 8.10^{10} p$^+$/cm^2). Window rate: $e_n = 274.6$ s^{-1} — : DLTS spectrum after irradiation at 77 K; ● : DLTS spectrum after annealing (15 min at 320 K); ↓ : upper temperature for the study of irradiation defects formed at 77 K by DLTS spectra

DISCUSSION

A) Analysis of the optical capture cross section $\sigma_n^O(h\nu)$ of E_3

We just give here results of the theoretical fit of the experimental curve $\sigma_n^O(h\nu)$ given by the Fig. 1. Chantre and Bois [14] have shown that $\sigma_n^O(h\nu)$ or $\sigma_p^O(h\nu)$ can be fitted by a theoretical model whose parameters are: the Franck-Condon shift $d_{FC}:E_n^O = E_n + d_{FC}$, the extension α^{-1} of the electron wave function on the trap, the relative weights of the transitions towards L and X minimum with respect to the transitions towards Γ minimum.

The theoretical $\sigma_n^O(h\nu)$ corresponding to: $E_n^O = 0.44$ eV, $d_{FC} \approx 0.2$ eV, $\alpha^{-1} = 4.5$ Å, $P_L/P_\Gamma = 0$, $P_X/P_\Gamma = 0.06$ is given in Fig. 1. We note a good agreement between experimental and theoretical curve. Fig. 6 describes the E_3 defect in a configuration-coordinate diagram with the thermal and optical energies determined by DLTS and DLOS measurements.

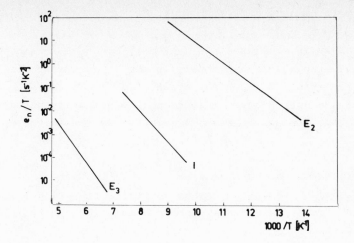

Fig. 5. "Signatures" of the defects created at 77 K (temperature window: 77 K - 200 K)

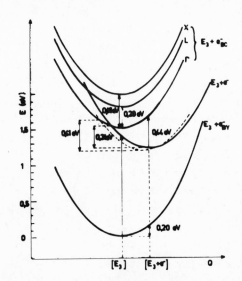

Fig. 6. Configuration coordinate diagram for the E_3 defect: electronic and elastic energy as a function of the configuration coordinate Q.

B) Irradiation defects created at 300 K

Electron or proton irradiations (at low doses $\leq 5 \ 10^{11} \ p^+/cm^2$) at 300 K create the same defects with two main differences:

i) the creation yields of the E_4 and E_5 defects with respect to that of E_3 are 5 times more important for proton irradiations,

ii) the D_1 defect does not appear in the case of electron irradiation. The defect could be of a new type and due to proton implantation:

one can expect heavier particles to create more complex damage than electrons.

The result i) could be interpreted if E_4 and E_5 are more complex defects than E_3 but this hypothesis is not in agreement with the following results: - the concentration of these defects increases linearly with electron dose independently of impurity concentration [6] - they have the same threshold energy creation as the E_2 and E_3 defects.

C) Irradiation defects created at 77 K

The E_2 and E_3 traps are created by a low temperature irradiation. They are not formed by secondary reactions during the annihilation stages I and II. This important result is a new proof that E_2 and E_3 are certainly simple defects in agreement with recent hypothesis about their nature: a vacancy for E_2 [9] and As_{Ga} for E_3 [17].

Following Thommen's and Pons' results [12,9] on the variations of the introduction rate of the defects as a function of energy, it seems that the I centre could be the di-vacancy, but this conclusion is rather speculative and only the study of its introduction rate as a function of energy could give a definitive conclusion.

For the moment, we cannot affirm that the increase of $\Delta C/C$ for E_2 and E_3 (Fig. 4) after an annealing at 300 K is significative of an increase in the concentration of E_2 and E_3 or that it is only due to the decrease of the depletion depth because of the decrease of compensation. However, if the corrections due to the variation of the depletion depth are legitimate, then the concentration of E_2 would increase by 30 %, while that of E_3 would remain constant after the room temperature annealing.

Therefore the increase of E_2 concentration would be nearly equal to the number of I centers which thermally anneal and this result would be in agreement with the hypothesis that it would be the di--vacancy and E_2 a vacancy, because the di-vacancy would anneal during stages I or II by recombination with a nearly intersititial defect (As_i or Ga_i) that would give:

- either a simple vacancy (V_{Ga} or V_{As}), therefore an increase of E_2,

- or a simple vacancy associated with an antisite defect ($V_{As}+As_{Ga}$ or $V_{Ga} + Ga_{As}$).

ACKNOWLEDGEMENTS - The authors gratefully acknowledge G.M.Martion of L.E.P. for the VPE samples, D.Pons, J.C.Bourgoin for electron irradiation and the Organisation D.G.R.S.T. for its financial support.

REFERENCES

1. D.V.Lang, R.A.Logan, J.Electron.Mat. $\underline{4}$, 1053 (1975)
2. A.Mitonneau, G.M.Martin, A.Mircea, Electron.Letters $\underline{13}$, 666 (1977)
3. G.M.Martin, A.Mitonneau, A.Mircea, Electron.Letters $\overline{13}$, 191 (1977)
4. A.Mircea, D.Bois, Int.Conf. on Rad.Effects in Semicond., Nice (1978)
5. D.V.Lang, L.C.Kimmerling, Lattice Defects in Semicond. 1974 (Inst. Phys.Conf.Ser. 23, p.581, 1975)
6. D.V.Lang, Rad. Effects in Semicond. 1976 (Inst.Phys.Conf. Ser.31. p.70, 1977)
7. D.V.Lang, R.A.Logan, L.C.Kimmerling, Phys.Rev.B. $\underline{15}$, 10, 4874 (1977)
8. D.V.Lang, C.H.Henry, Phys.Rev.Letters $\underline{35}$, 22, 1525 (1975)
9. D.Pons, Thèse Docteur-Ingénieur, Université de Paris (1979)
10. D.Pons, A.Mircea, A.Mitonneau, G.M.Martin, Int. Conf. on Rad. Effects in Semicond., Nice (1978)
11. D.Pons, P.M.Mooney, J.C.Bourgoin, à paraître dans J.A.P.
12. K.Thommen, Rad. Effects $\underline{2}$, 201 (1970)
13. A.Chantre, Thèse de Docteur-Ingénieur, Université de Grenoble (1979)
14. D.Bois, A.Chantre, Conf. SFP Toulouse (1979). To be appear in J.Phys.
15. The stopping and ranges of ions in matter: Vol.3, ed. J.F.Ziegler, Pergamon Press (1977)
16. H.Matsumura, M.Nagatomo, S.Furukawa, K.G.Stephesn, Rad. Effects $\underline{13}$, 121 (1977)
17. A.Zylbersztejn, Int. Conf. on Deep Levels in Semiconductors, Ste Maxime (1979)

PROPERTIES OF AN EXTENDED DEFECT IN GaAs$_{.62}$P$_{.38}$

George Ferenczi

Research Institute for Technical Physics
H-1325 Budapest, Ujpest 1, P.O.Box 76.
Hungary

Properties of a common trap in device grade GaAs$_{.62}$P$_{.38}$ have been analysed. DLTS and single shot capacitance transient measurements revealed that the trap, which acts as an effective recombination center, has collective properties i.e. a single time content Shockley--Read description is not applicable. The trap is identified as line defect.

It became a common belief in recent years that the study of native deep states of optoelectronic materials can lead: a) to the characterization of the minority carrier lifetime limiting non-radiative processes, b) to the understanding of the device degradation mechanism. These practically and theoretically important objectives were pursued recently by many authors [see e.g. 1-5]. One level which presence is usually reported in about 0.4 eV below the conduction band. Over the years, studying the properties of GaAs$_{.62}$P$_{.38}$ VPE layers and p-n junctions we found this level in every sample from 5 different manufacturers The techniques, used for detecting this level were: TSC, Admittance Spectroscopy, TSCAP, DLTS and finally single shot capacitance transient measurements. In their recent paper Lopez et al. [5] report three electron trap at 0.02, 0.38 at 0.56 eV from the conduction band. This paper intends to prove that these levels belong to the same extended defect and there is no way to single out individual activation energies.

EXPERIMENTS

The measurements were carried out on Au Schottky barriers fabricated on n-type VPE GaAs$_{.62}$P$_{.38}$ epitaxial layers. The net donor concentration varied from 5×10^{16} cm^{-3} to 5×10^{18} cm^{-3}. The layers were doped with Se, Te, S or Zn. To study hole injection, occasionally p^{+}n diodes were prepared by Zn difussion or commercially available red LED's were used. The DLTS implementation used the lock-in variety [6]. The capacitance transient measurements were carried out using a digital transient recorder with 10 μs minimum sampling time (System KFKI).

RESULTS

A typical DLTS run is shown on Fig. 1. Three feature are obvious from these curves:

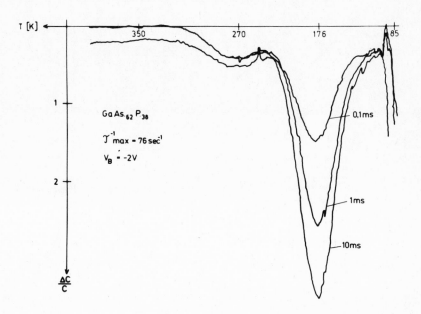

Fig. 1. DLTS runs at three different majority pulse width on a GaAs $_{.62}$P $_{.38}$ p$^+$n junction

i) There is always a side minima at 270 K.

ii) The line shape and the half width of the DLTS peak is well defined for the case of lock-in averaging [7]. The present peak is broader than expected indicating non-exponential decay.

iii) Applying different majority pulse widths, capture rate measurement was attempted. The strong non-exponentiality is obvious. Detailed results of the capture rate measurements for three different temperature are shown on Fig. 2.

The Arrhenius plot of the DLTS peak temperature for measurements made at different rate windows gives an uncorrected activation energy of 0.38 eV, however observation ii) and iii) points toward the possible non-exponentiality of the emission process. This is indeed the case, Fig. 3. The suprisingly strong non-exponentiality should be a warning for the users of the DLTS technique. This type of signal processing chooses the leading Fourier component hence a non-exponential decay containing a strong exponential factor as well will be

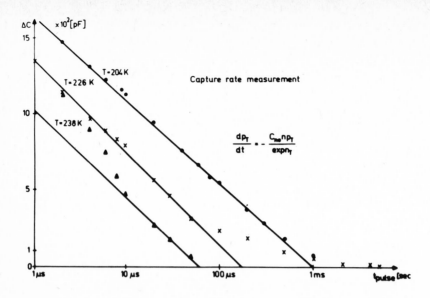

Fig. 2. Capture rate measurement at three different temperature on the same p⁺n junction. The measurement was made via DLTS set up. The results are corrected for the edge effect [8].

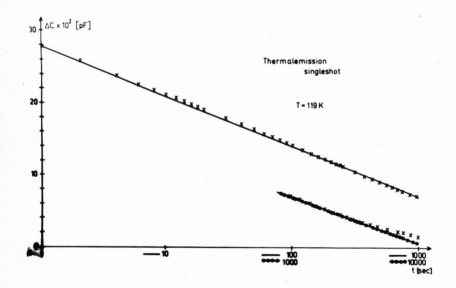

Fig. 3. Single shot capacitance transient measurement on the same diode.

transformed by the DLTS processing very similarly to a truly exponential decay. Obviously the measured concentration is quite erroneous in this case.

The non-exponentiality may be the result of a strong electric field dependence [9]. A trap profiling experiment should be decisive in this respect. We do know from separate experiments on epi-layers, that the trap distribution is homogeneous. So a profiling experiment, where the trap emission comes from varying electric field regime, should display the field dependence. This is not the case (Fig. 4).

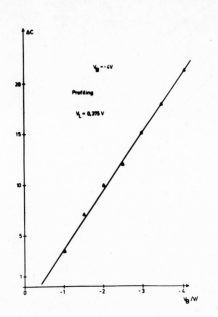

Fig. 4. The trap profiling shows no field dependence.

The thermal emission measurement shown on Fig. 3 is not satisfactory in one respect. The result seems to indicate an infinite concentration for the trap (an extrapolation for infinitesimally short time after the majority pulse). By using a digital transient recorder we get to 10 μs from the bias pulse and extended the time base for the transient recording for six order of magnitude. The results for three different temperature are to be seen on Fig. 5. These curves, however untypical, have at least meaningful asymptotic behaviour.

One way to explain the result is to assume correlation effect among the captured carriers. By capturing increasing number of carriers a repulsive electric field starts to build up and the capture becames increasingly difficult. The reverse process is valid for emission.

With other words: single time constant Shockley-Read statistic is invalid, the capture and emission rate depend on the number of the already captured carriers. The following differential equation is suggested:

$$\frac{dp_{\tau}}{dt} = -\frac{c_n n_o p_{\tau}}{\exp(1-p_{\tau})} \tag{1}$$

Eq.(1) cannot be solved analytically. Fig. 6 shows the numerical solution of the rate equation. The similarity to the measured data (Figs. 2 and 5) are obvious.

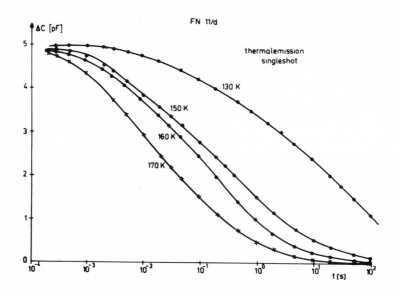

Fig. 5. Single shot capacitance transient measurement at three different temperatures. Measurement made on an Au-n GaAsP Schottky diode.

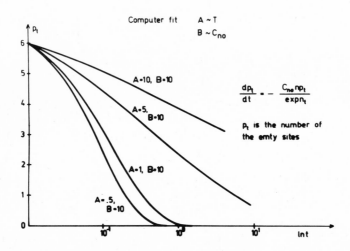

Fig. 6. Numerical solution of the proposed rate equation for different A,B parameter values.

DISCUSSION AND CONCLUSIONS

The assymptotic solution of Eq.(1) suggests that for very low occupation numbers, the site correlation effect is not longer active, i.e. the thermal emission decays should have an exponential tail. This is indeed the case. From the exponential tail the ground state energy of the extended defect can be determined. The measured value is 0.1 eV. The next question is the identity of the observed extended defect. Eq.(1) can be easily understand by assuming the build up of a repulsive Coulombic barrier. Such a configuration is produced by a line defect i.e. a dislocation having unsaturated dangling bonds. To check this we measured the number of non-radiative dark spots on the C.L. image of an S.E.M. The dark spot concentration and the trap concentration correlated over two orders of magnitude with the trap concentration. This seems to suggest that the so called 0.38 eV trap in $GaAs_{.62}P_{.38}$ belongs to dislocations with a non defined activation energy. Minority carrier injection experiments revealed:

The trap is a recombination center having about the same electron and hole capture rates.

To conclude we suggest that the electron trap present in all VPE $GaAs_{.62}P_{.38}$ material can be identified as a dislocation. This dislocation acts as a non-radiative recombination center.

Dislocations as non-radiative recombination spots were identified several times [e.g. 10], but we believe this is the first report on their direct investigation by space-charge spectroscopy, a tool which enables the direct measurement of their influence and control on the minority carrier lifetime.

ACKNOWLEDGEMENT - A part of this work has been accomplished in Lund. The hospitality of the Royal Swedish Academy is acknowledged. Special thanks due to H.G.Grimmeiss and L-Å.Ledebo for the helpful discussions and valuable suggestions. The valuable assistence of J.Boda, J.Kiss and M.Somogyi was a real help.

* Note added in proof. After the conference, the authors attention was drawn to the work of J.R.Patel et al. [11]. Using DLTS they did detect a series of states, attributed to deliberately produced dislocation. It is interesting to note their Fig. 2 where a typical non-exponenti-al capture process is revealed. Unfortunately they did not comment on this observation. The pioneering work of T.Figielski [12] should also be mentioned. Their photoconductivity decay measurements on deformed Si and Ge did also show the non-exponential behaviour.

REFERENCES

1. H.Schade, C.J.Nuese and J.J.Gannon, J.Appl.Phys. $\underline{42}$, 5072 (1971)
2. L.Forbes, Solid-St. Electron. $\underline{18}$, 635 (1975)
3. S.Metz and W.Fritz, Solid-St. Electron. $\underline{20}$, 603 (1977)
4. C.López, A.Garcia, F.Garcia and E.Mûnoz, Solid-St. Electron. $\underline{22}$, 81 (1979)
5. B.Tell and C. van Opdorp, J.Appl.Phys. $\underline{49}$, 2973 (1978)
6. L.C.Kimerling, IEEE Trans.Nucl.Sci. NS-$\overline{23}$ (1976)
7. G.L.Miller, D.V.Lang and L.C.Kimerling, Ann.Rev.Mater.Sci. 377 (1977)
8. L.C.Kimerling, J.Appl.Phys. $\underline{45}$, 1839 (1974)
9. D.Pons and S.Makram-Ebeid, to be published in J. de Physique
10. A.Rasul and S.M.Davidson, Inst.Phys.Conf.Ser. $\underline{No.33a}$, 306 (1977)
11. J.R.Patel and L.C.Kimerling, J. de Physique, $\overline{C6\ 40}$, C6-67 (1979)
12. T.Figielski, Solid-St. Electron. $\underline{21}$, 1403 (1978)

LARGE DEFECT-LATTICE RELAXATION PHENOMENA IN SOLIDS

Jerzy M. Langer
Institute of Physics,
Polish Academy of Sciences
02-668 Warsaw, Al.Lotników 32/46, Poland

ABSTRACT

The talk gives a brief introduction to the large lattice relaxation (LLR) phenomena in solids with main emphasis given to the deep levels in semiconductors, and effects caused by a charge exchange between defects and bands (ionization and capture). Several experimental examples serving as the evidence for LLR are presented and discussed.

INTRODUCTION

In a semiconductor physics the one-electron picture is adequate in description of the most phemomena observed. Moreover, lattice vibrations can be incorporated in a form of phonons more or less affecting electronic properties of solids. Such a simplified picture fails while describing properties of localized objects in crystals. The so called "deep levels" can serve as a good example of this breakdown. For a long time, defects, and especially impurities, in solids have been described using only the electronic picture, totally neglecting any involvement of a defect-lattice coupling. For a group of the effective-mass impurities, whose electronic (or hole) wave function extends over many unit cells, the coupling of electronic and lattice motion is really weak. Going to more localized defects this interaction must, and indeed plays an increasing role irrespectively of the ionicity of the solid. It is the aim of this lecture to show and prove how important is the defect-lattice interaction and to how bizarre effects it can lead.

Recently several anomalies in the properties of the deep levels in semiconductors have been observed, indicating breakdown of a simple one-electron picture of these impurities or defects. Perhaps the most spectacular is the observation of the metastable effects [1-18]. A typical example is a persistent photoconductivity shown in Fig. 1 for the so called DX center in Te doped GaAlAs crystals [4,5].

The defect states exhibiting the memory effects usually have large optical ionization energy (sometimes even larger than a bandgap as e.g. for "O" center in GaAs [10]) much larger than thermal, and the

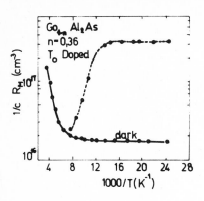

Fig. 1. Persistent photoinduced conductivity in n-type $Ga_{0.64}Al_{0.36}As:Te$ (---) compared with the carrier concentration measured for the sample cooled in the dark (——) [4]. At $T < 50$ K lifetime of the photoinduced state becomes immeasurably long.

metastable effects are caused by a potential barrier separating these anomalous impurity states from "normal" impurity (or band) states. The charge capture is thermally activated and its cross section is usually very small ($< 10^{-30} cm^2$ [3,5]). The most surprising is that such states can be found in almost every semiconductor from almost purely covalent InSb [6] to a typical ionic semiconductor CdF_2 [9].

For some time such bizarre states have been classified as the impurity states connected with subsidiary minima of the conduction band [2,8,13], and long lifetime explained as due to small coupling between different minima. However, the pressure coefficient of some of such states has been shown to be larger than those of the bands [7,13] and the inter-valley interaction matrix element is much too large to explain the metastable level occupancy, which can last even for days.

Meanwhile a concept of the multicharged defects has been advanced [1,2]. Again this notion can hardly be accepted in a view of a recent capacitance [5] and mobility [4,14] measurements. In 1974 [15,6] Vul and Shik proposed a model of the inhomogenous semiconductor with internal barriers between the high and low resistivity regions. In spite of several qualitative and quantitative successes [17] the model seems to be hardly acceptable for strongly doped semiconductors, and what is most important, its quantitative predictions must strongly depend on concentration, what is in some cases contradictory with the experiment [18].

All the three models based upon purely electronic treatment of the doped crystal, totally neglecting coupling of the levels with lattice vibrations. This reflected attitude of the most researches, especially those working with covalent compounds, that vibronic effects are either negligible or very small. Indeed, most experimental data concerning both shallow and deep impurities [19,20] supported this view. Incorporation of the strong impurity-lattice coupling in a form of e.g. configuration coordinate model [3-7, 9-12, 14, 18, 21] resolves all the inconsistencies mentioned above, and gives simple and natural explanation of all the metastable effects observed.

From the growing number of experimental data, which I am going to discuss briefly below, it follows that defect states exhibiting strong coupling to the lattice must be treated as normal defects, and not as a scientific curiosity. Moreover, acting as traps, these defects can strongly affect temporal characteristic of the solid-state electronic devices [4]. In a most extreme case nonradiative recombination of the carriers at such defects causes their motion [22]. It should be pointed out, however, that their proper theoretical description is almost totally lacking. Most of the proposed models are nothing more than a simple phenomenology. Necessity of a non perturbational treatment of the impurity-lattice interaction, and possible breakdown of an adiabatic approach, are strong obstacles in any serious attempt of a model calculation of these defects. Moreover, large lattice relaxation effects should have additional confirmation in such experiments as EPR or ENDOR not only to give some proof or a lattice rearrangement, but especially to give information about actual defect symmetry and its electron distribution. To my knowledge, there are no such measurements performed until now.

In this paper I present some experimental evidences for LLR and give a general characteristics of the processes occuring in this class of defects. Interconnection between phenomena observed in a highly ionic solids and normal semiconductors is stressed. In a final part some remarks relevant to a future research in this field are given.

STRONG DEFECT-LATTICE COUPLING
THE CONFIGURATION COORDINATE MODEL AND ITS CONSEQUENCES

Large lattice relaxation (LLR) effects are very common in highly ionic solids. Due to strong electrostatic interactions, almost any electronic excitation must be accompanied by the lattice polarization effects. The best known is a coupling of a band carrier (electron or hole) with the lattice vibrations leading to a formation of a new quasi-particle called polaron. Dependent on the strength of the long range electrostatic interactions (due to LO phonons) and the short range (deformation potential or molecular-like bonding), the radius of the polaron can be either large (large polaron, whose effective mass is at most 2-3 times a "bare" carrier effective mass) or small (small polaron behaving as a sort of a molecular center, e.g. self trapped hole in alkali halides). Criteria for either large or small polaron formation

have been a subject of several studies by Toyozawa [23-25] and Emin
with Holstein [26].

One of the best methods to observe and study coupling of the
electronic excitations to the lattice and the LLR effects is the optical spectroscopy. A classical example is the F centre [27], being an
electron trapped by a positive potential of an anionic vacancy.
Absorption and emission spectra corresponding to the same electronic
excitations are shifted by one or more eV (Fig. 2a). Such a large
Stokes shift and Gaussian broadening of the optical transitions are
caused by a dramatic change of the electron wave-function localization
when exciting from a ground (localized 1s state) to any of the excited
states of the defect (delocalized, effective mass states) [28]. Fig.2b
shows absorption spectrum of the F-centers in the relaxed excited

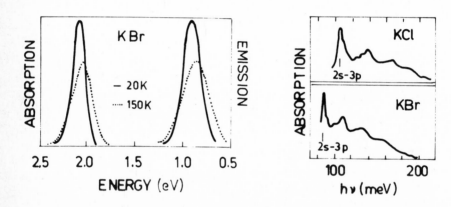

Fig. 2. a/ Absorption and emission of the F-center in KBr crystal
        b/ Relaxed excited state absorption spectrum of the F-centers in KCl and
           KBr crystals [29].

states [29]. Note more than ten-fold energy scale difference between
1s-2p and the next higher excited states. This is a manifestation of
an electron delocalization in the excited states of this center. This
delocalization results in a drastic change of a local charge screening
and causes very large (almost 10%, as was calculated by Wood and coworkers [30,31]) local lattice expansion. Similar effect is expected
for   ionization of a localized defect.

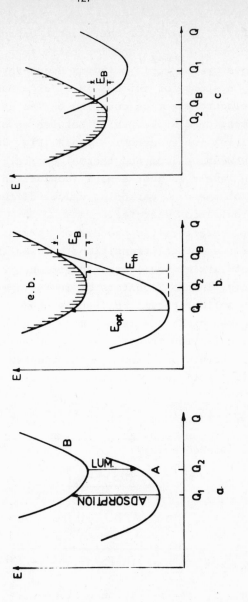

Fig. 3. Typical configuration coordinate diagrams in solids E is a sum of electronic and elastic energies, Q is a configuration coordinate.

Strong coupling of the defect with the lattice vibrations is best visualised using the configuration coordinate diagram (Fig. 3). A total energy of the system E (electronic plus elastic, usually taken in a harmonic approximation) is plotted versus lattice displacement represented by the so called configuration coordinate Q. In most cases this coordinate represents some vibronic mode of the system and has no simple visualisation [32]. Change of the elastic energy of the system is called a relaxation energy E_{relax}, and its ratio to a value of characteristic vibronic frequency $\hbar\omega_{vib}$ is a famous Huang-Rhys S-factor. We shall consider only systems for which S is a large number (more than 10).

Strong electron-lattice coupling causes several effects. The first is the just mentioned Stokes shift between absorption and luminescence spectra involving two electronic levels (Fig. 3a). An equivalent is a difference between optical and thermal excitation energies (Fig. 3b). If one of the levels is just a band (conduction or valence), horizontal displacement of the c.c. parabolas causes difference between optical and thermal ionization energies.

The next characteristic phenomenon is vibrationally induced broadening of the optical transitions between the levels characterized by a different lattice equilibrium position [32]. Intradefect transitions (as e.g. in the F centers or in the spectra of transition metal impurities in solids) became bell-shape bands with all the internal structure smeared with increasing value of a Huang-Rhys S-factor. In a case of photoionization transitions the observed spectrum is a convolution of an electronic and vibronic components, and is given by a very simple formula (valid when S is large) [33,9,21]:

$$\alpha(\hbar\omega) = \frac{1}{\sqrt{\pi}} \int_{-\beta}^{\infty} \alpha_{el}(E_{opt}, \hbar\omega + \Gamma z)(1 + \frac{\Gamma z}{\hbar\omega})e^{-z^2}dz, \qquad (1)$$

where β is a reduced energy

$$\beta = \frac{\hbar\omega - E_{opt}}{\Gamma} \qquad (2)$$

and broadening Γ is:

$$\Gamma = (\frac{\omega_{exc}}{\omega_o}) \sqrt{2(E_{opt} - E_{th})\hbar\omega_o coth(\hbar\omega_o/2kT)} \qquad (3)$$

Here ω_{exc} and ω_o are the vibronic frequencies of a lattice in the both involved states. One of the first examples in a semiconductor impurity physics, for which such a broadening have been observed was O_P in GaP [34,35] (Fig. 4).

Impurity lattice coupling affects in a same way the reverse process: carrier capture by the localized defects. If the coupling was none or very small the carrier capture cross section would have been almost temperature independent. It is not true in the reverse case. Carrier capture occurs most effectively if the accepting level enters into the band states (vicinity of $Q = Q_B$ at Fig. 3b) due to lattice

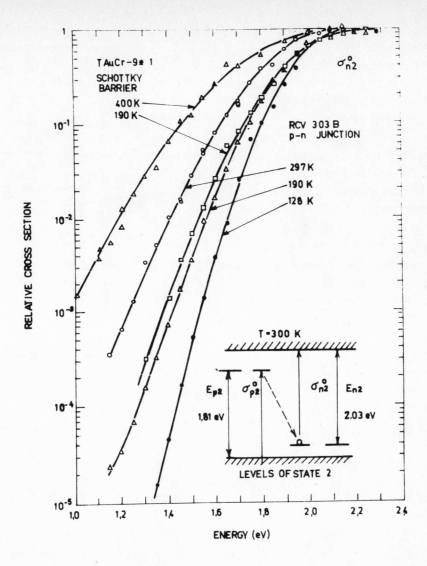

Fig. 4.   Influence of temperature on a vibrationally broadened photo-
          ionization spectrum of a two-electron O_p state in GaP [34,35]

vibrations [35]. Since this process requires overcoming of a potential
barrier E_B, it must be temperature dependent. In a high temperature
limit capture cross section is simply thermall activated
$\sigma = \sigma_\infty \quad \exp(-E_{act}/kT)$, and $E_{act} \approx E_B$. This is the so called Mott
limit. Fig. 5 shows several experimental examples of temperature de-
pendent capture cross sections for different defect levels in GaP and
GaAs crystals [35]. During carrier capture energy is dissipated in a

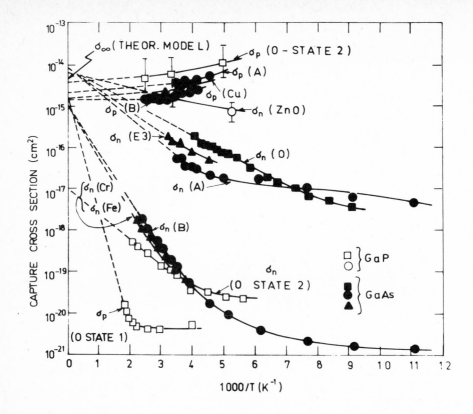

Fig. 5. The temperature dependent electronic σ_n and hole σ_p capture cross-section for several traps in GaP and GaAs crystals exhibiting LLR effects.

form of violent local vibrations, so the process is called MPE – multiple phonon emission, and is now generally accepted as dominant for deep levels in semiconductors.

Relative positions of the C.C. parabolas is a crucial parameter leading either to normal (Fig. 3b) or metastable cases (Fig. 3c). In the second case at low temperatures metastable occupancy of the excited states is possible due to the presence of a barrier E_B at Q_B. Dependent on the nature of a barrier it can be overcomed either by a thermal excitation over its top or via tunneling. If a single C.C. diagram is a good description of a system this distinction is meaningless. Tunneling is a normal process, but in a very high temperature limit it can be approximated by the thermal excitation over the barrier. It seems, however that in most physical cases this limit (called the Mott activation model) is not reached [36,21,37]. Therefore

the observed activation energy must be smaller than the true geometrical barrier height and should be temperature dependent [38]. Figure 6
can serve as quantitative illustration of this point. Although parameters used are pertinent to the CdF_2:In case [21], a general validity
of the conclusion is preserved. In the more complicated cases (see
discussion of the CdTe:Cl below), while a single C.C. diagram is not
adequate and especially if a system undergoes really large rearrangement (e.g. interstitial jumps), tunneling due to a large distance to
be travelled is less effective and the simple Mott activation model
($E_{act} = E_{barrier}$) can be valid.

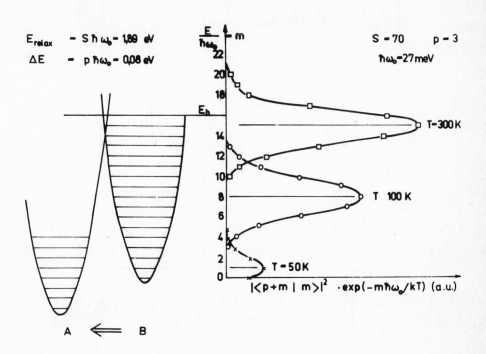

Fig. 6. Quantum-mechanical tunneling through a barrier separating two
C.C. parabolas. Right portion shows relative contribution of
different tunneling levels at the tree temperatures.

Coupling of electronic transitions to the lattice vibrations is known to quench diagonal electronic interaction matrix elements. This is known as the Ham effect [39], quite common in all the Jahn-Teller phenomena. Here I would like to point out another interesting feature related to the Ham effect: quenching of the Fano effects. It is known that interaction between discrete state degenerate with a continuum of the band states leads to a dispersion-like shape of a discrete-discrete state optical transitions (the Fano effect) [40-42] (Fig. 7b). If the

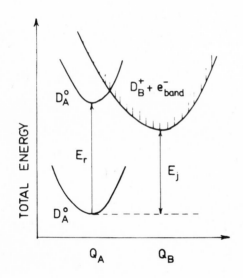

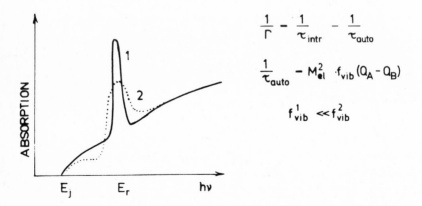

$$\frac{1}{\Gamma} - \frac{1}{\tau_{intr}} - \frac{1}{\tau_{auto}}$$

$$\frac{1}{\tau_{auto}} - M_{el}^2 \cdot f_{vib}(Q_A - Q_B)$$

$$f_{vib}^1 \ll f_{vib}^2$$

Fig. 7. Quenching of the Fano effects by LLR
a) C.C. diagram of a resonant excited state of a defect;
b) spectral profile of a resonance transition on a photo-ionization background without (1) and with (2) auto-ionization modification.

lattice equilibrium positions for the resonant and band states are different (Fig. 7a), interaction matrix element between the band and resonant states will be quenched (analogue of the Ham effect), and all the antiresonance features must be less pronounced or even not obesrvable (Fig. 7b). Moreover, an autoionization lifetime should increase because of the same reason. Its unquenched value is of the order of 10^{-13}s or even shorter [40]. Experimental example of such a quenching was found in case of Co doped II-VI compounds. From the photoconductivity measurements it has been shown that some of the excited states of the transition metal impurities are resonant [43,44]. Very large negative dips in a photocurrent spectrum of CdSe:Co (Fig. 8) are explained as being result of much more rapid internal nonradiative recombination rate than autoionization lifetime. Analyzing other experiments relevant to this effect, Radlinski [44] have shown that auto-

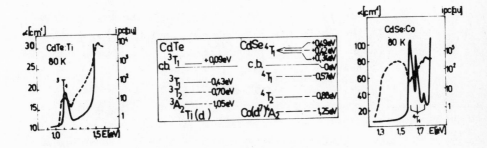

Fig. 8. Photoconductivity enhancement (Ti) or quenching (Co) by excited states of transition metal impurities degenerate with the conduction band [43].

ionization life-time is longer than 10^{-8}s. This can be considered as a good example of quenching of the Fano effect by LLR. Additional support for this interpretation comes from experiments on photoionization of

Cr in the II-VI compounds, where it has been shown that the lattice distortion [45,46] during photoionization transition is much larger than for the internal Jahn-Teller distortions [46]. Another example of total quenching of autoionization process will be given while discussing metastable Cl states in CdTe crystals.

ANOMALOUS IMPURITY STATES EXHIBITING LLR EFFECTS

Amont the most spectacular effects caused by LLR is occurance of metastable effects. They are caused by a barrier separating the two coupled electron-vibronic states. Below I present three experimental examples of defects exhibiting metastable effects. I have chosen them to show surprising similarities of the effects in crystals of drastically different ionicities and defects of different origin. The two other: DX centers in GaAlAs [4,5] and "O" centers in GaAs [10-12] have been mentioned in the introcution.

$$CdF_2 : In \quad [9,21]$$

Indium impurity has been shown [9] to possess two neutral In^{2+} states differing in their optical properties due to different electron wave-function localization. A shallower is a typical for CdF_2 delocalized hydrogenic-like donor [48], whose ionization energy is about 0.1 eV. A deeper (its thermal ionization energy is about 0.25 eV) is interpreted as a localized, atomic-like In^{2+} state. Its optical ionization energy, as determined from the photoconductivity spectrum (Fig. 9), and than more precisely from a shape of a photoionization absorption (Fig. 10), is about 1.9 eV. The enormously large difference between E_{opt} and E_{th} (about 1.7 eV' indicates very strong impurity-lattice coupling. At low temperatures only deeper In_{loc}^{2+} state is populated, but complete metastable occupancy inversion can be achieved by illumination of a sample by whie or UV light. At temperatures below 50 K lifetime of the inverted state is immeasurably long. Decay of the metastable occupancy, monitored either by observation of the both In_{loc}^{2+} and In_{hydr}^{2+} absorptions, or magnitude of persistent photoconductivity [21], occurs via tunneling thorugh a barrier separating the two states (Fig. 11). The activation energy is smaller than geometrical barrier height ($E_{act} \approx 0.6\ E_B$) [21,47], as expected for the tunneling process, discussed previously. In accordance with a C.C. model, photoionization transitions of the localized In^{2+} state are vibronically broadened and their shape $\alpha\ (\hbar\omega)$ and temperature dependent broadening $\Gamma(T)$ are well described by a

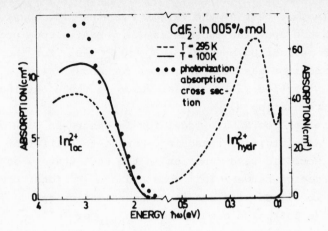

Fig. 9. Optical spectrum of CdF$_2$:In crystals [9]. Absorption measurements were made without broad-band illumination.

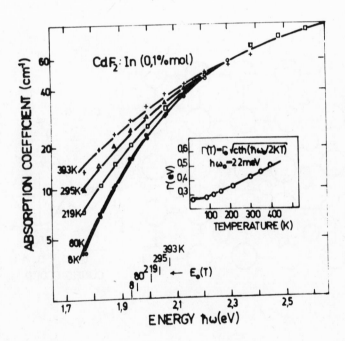

Fig. 10. Vibronically broadened photoionization absorption of the localized In^{2+} state. Solid curves are calculated using Eq.1. Insert shows the temperature dependence of the broadening parameter Γ(T) (solid line gives fit to Eq.3) [9,21].

formulae (1-3). Another test of the validity of the C.C. model is the influence of pressure upon the two states [21] . Basing upon pressure dependence of the relaxation energy ($E_{relax} = E_{opt}-E_{th}$ for the In^{2+}_{loc} state), local distortion has been estimated for about 8 % of a n.n. distance in a CdF_2 lattice [21], similar value to those obtained for excited F-centre relaxation [30].

The configuration coordinate model for In centre in CdF_2 crystals has quite simple and intuitive reasoning. Due to the different localization of the electronic wave-function in the both states, screening of Coulombic attraction between In^{3+} core and n.n. F^- ions, present when electron is localized within a first coordination sphere (In^{2+}_{loc}), is canceled after electron delocalization (In^{2+}_{hydr} or $In^{3+}+e^-_{band}$). It must be accompanied, similarly as in case of F centers, by a large local lattice collapse (Fig. 11). Here a configuration coordinate can be interpreted as an effective n.n. distance, and interaction mode is just a breathing mode.

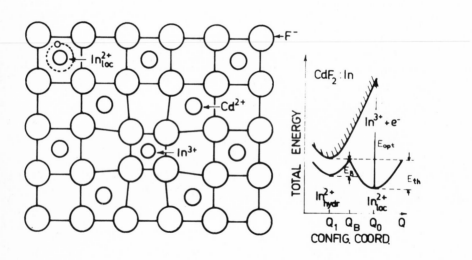

Fig. 11.   Microscopic model of the LLR effect in CdF_2:In. The configuration coordinate diagram of the In impurity is shown in the insert

Extrinsic InSb [6,49,50]

A model of a local symmetrical distortion of the lattice around a charged defect has a rather sound justification in highly ionic solids. Going to more covalent crystals, electrostatic effects are less pronounced and it would be rather surprising if similar to CdF_2:In metastable single impurity states were found. In contrary, involvement of some sort of complex or even large defect motion would be more probable source of metastable states if any. A good example of such a defect have been found by Porowski and coworkers in an extrinsic InSb [6,49]. Its exact chemical indentity is not clear. Some analogies with GaSb:Te, O, S [8] would suggest trace contamination by oxygen. Anyhow, the metastable effect found in InSb are very similar to those just described in CdF_2:In case.

Due to a very small band-gap, most information on this system came from pressure dependent thermogalvanomagnetic experiments. The main results are summarized in Fig. 12. At the ambient pressure carrier concentration is almost temperature independent (Fig. 12a). Drastic changes are observed when pressure exceeds 8 kbars. As it is clear from Fig. 12a, two different levels govern temperature dependence of the carrier concentration. Most surprising is dicontinuity at $T_c \approx 130^o$K. At lower temperatures only shallower D_A level is active. The discontinuity and hysteresis around T_c indicate strong temperature dependence of a carrier capture cross section of the level D_B, indicating its strong coupling to the lattice. Metastable occupancy of a level B are best documented in Fig. 12b. If there is no pressure applied during cool-down cycle to 77 K, at low temperature only level A is active in a pressure dependence of carrier concentration. In contrary if the pressure is high, all electrons are trapped by level B at low temperatures even after pressure release (then both levels are again resonant - see Fig. 12c). Therefore as it is shown by a bottom curve of Fig.12b, level A is not active and the carrier concentration (due to background doping) is pressure independent.

From the kinetics measurements [49] the barrier separating the two levels has been found to be very high (about 0.3 eV). As explanation of all these bizarre effects, Porowski et.al. [6] proposed a model of a defect exhibiting very large lattice rearrangement (Fig. 12d). The essence of the model has been possibility of a defect to move in a lattice from A to B positions characterized by different lattice energies E_A and E_B and electron energies ϵ_A and ϵ_B. The barrier is mostly due to a lattice motion. The model proposed is just an extended

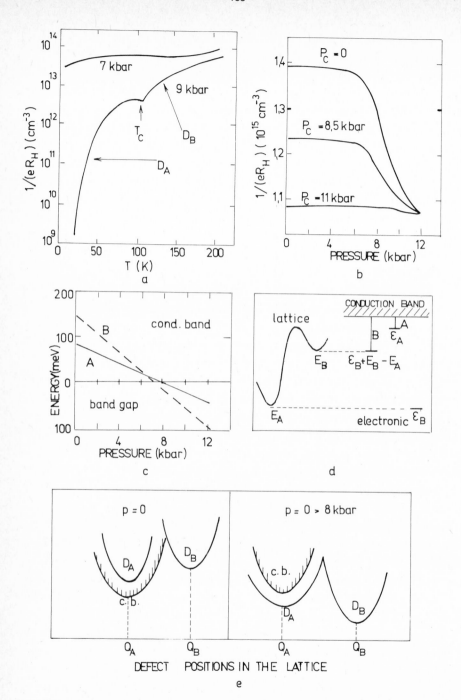

Fig. 12. Anomalous states in extrinsic InSb. a) influence of pressure on temperature dependent electron concentration [6]. T_C indicates region of hysteresis and time transients; b) pressure dependence of electron concentration at 77 K. P_C are pressures during cool-down cycle [6]; c) pressure dependence of energetic positions of normal (A) and anomalous (B) states [6]; d) two-position model of anomalous center in extrinsic InSb 6 ; e) c.c. model of the center at different pressures.

version of a c.c. diagram (possibility of two dimensional motion: translation + local vibrations). A simplified version is plotted in Fig. 12e. Since possibility of large lattice rearrangement is high (eg. vacancy motion or so) the single c.c. diagram description might to inadequate.

It is worthwhile noting that historically it has been the second after Wright et.al. [3] case, where the authors suggested and morover presented sound arguments for a lattice rearrangement at such a defect, being the source of metastability effects.

Very similar behaviour have been observed for GaSb:Se, Te, O [8], and again it seems [18] that a lattice rearrangement is a governing factor of the defect properties. It should be noted, however that an alternative model of an inhomogeneous semiconductor possesing internal $n-n^+$ barriers, responsible for bizarre kinetics, was proposed by Vul and Shik [15-17]. The model although generally correct cannot explain very large stokes shifts and magnitude of the observed effects [18].

$$Cd_{1-x}Zn_xTe : Cl \quad [7,14,51,52]$$

The properties of anomalous defect in InSb suggest a real motion of a defect (not a relatively small geometrically although energetically very large local distortion as in the case of CdF_2:In), and perhaps more complex nature of the defect. Another very much elaborated example supporting this point of view is CdTe:Cl, I have chosen this defect since the characteristic parameters of the defect involved $(E_{barrier}, E_{opt}, E_{th})$ indicate breakdown of a simplified single c.c. model.

The galvanomagnetic properties of CdTe:Cl crystal under hydrostatic pressure [7,14,52] (or what it appeared being equivalent; in the mixed CdZnTe:Cl crystals [51]) are governed by the one level responsible for metastable effects. At a very low Cl concentration additional hydrogenic level has been found, but at the concentrations under consideration these levels are smeared into a band tail.

At the ambient pressure at CdTe, a "Cl" level is degenerate with a c.b. (Fig. 13a). Application of a hydrostatic pressure or increase of Zn content, move the level into a band gap. In spite of its resonant character, the level can be filled up with electrons during cool-down cycle under hydorstatic pressure. After pressure release the level is still occupied and can be emptied optically (Fig. 13b). Its optical ionization energy at p = 0 is about 1.0 eV, as was found by a fit of Eq(1) to the experimental data (Fig. 13b) [15]. At the low temperatures

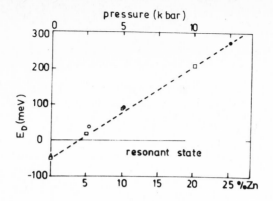

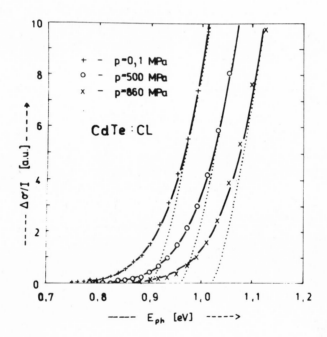

Fig. 13.   a) Energy position of the anomalous Cl level vs pressure or Zn content.
            Note that increase of Zn causes decrease of a lattice constant [51];
         b) Photoionization spectrum of a metastable Cl state at various pressures.
            Note that at lowest pressure the state is already resonant [14].

the life-time of this filled true resonant state can be infinite. This is the best illustration of the quenching of autoionization process (and the Fano effects) by a lattice relaxation, the effect I have discussed previously. Analysis of kinetics of a decay of pressure induced metastable occupancy of this level (or reversely its filling up at mixed crystals after light induced change of its occupancy) provides a value of a barrier height. It is one of the highest barriers ever found and its value is about 0.5 eV [7,14]. If a simple c.c. picture had been correct, and even the tunneling process had been neglected, enormous 1 : 4 ratio of the force constants of the defect before and after electron capture was implied. At ionic crystals [31] this ratio is large, but does not exceed a value of 1 : 2.

The explanation would be a large defect motion (e.g. between non-equivalent interstitial positions [52]. If it was true, the system had to be described by at least two-dimensional c.c. diagram (Fig. 14). In this case tunneling can be less effective and the Mott activation model would be more appropriate ($E_{act} \approx E_{barrier}$).

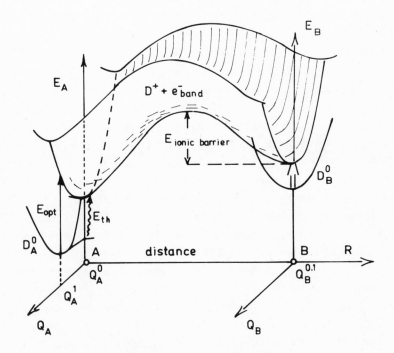

Fig. 14. Two dimensional configuration coordinate showing defect motion (R) as well as local lattice vibrations (Q_A) and (Q_B).

Finally I would like comment on designation of this level as the acceptor like. As a guideline, light induced change of mobility has been used [1,2,51]. Its increase would imply acceptor-like nature of the "Cl" defect. Such a designation is very much surprising due to clearly donor nature of Cl itself. If a defect was really an acceptor, its main constitutent had to be deep multicharged acceptor. The increase of mobility in this and other cases can be caused by different process. First if there were some compensation in a simple, light induced change of mobility due to change of ionized defect concentration would be rather small. Growth process [1,51] indicate high probability of such a compensation. In contrary, if electric transport was going through a band tail due to smeared density of hydrogenic-like donors, increase of mobility would have been just a consequence of motion of a quasi-Fermi level up into a conduction band. Strong support of this view is given by measurements of mobility dependence in CdTe:Cl [15] versus carrier concentration at constant temperature, and at ambient pressure. Different concentration are achieved by partial release of electrons from the metastable level. Therefore arguments on an acceptor-like behaviour of this and other similar defect must be treated with care.

DEFECT MOTION OR DEFECT PRODUCTION BY THE LLR

An extreme case of a strong impurity-lattice interaction is recombination induced defect motion or oven defect production. The basic mechanism of these processes is similar. During nonradiative recombination a very large portion of localized vibronic energy is emitted by a defect. This energy if not dissipated step by step can be a cause of large lattice rearrangement. Possibility of such a transformation is known for molecules, and can be treated along the formalism of the unimolecular reactions. This theory has been applied to solid by Weeks et.al. [53] and successfully explained experimental results of Lang and Kimmerling [54,55] indicating defect motion during nonradiative recombination.

At this kind of solid state reaction vibrational energy is gained from the recombination of the carriers. It must be localized to cause a lattice rearrangement, and a whole process resembles features of a local heating, since a motional barrier is overcomed by increase of the local vibronic energy. It is possible that purely electronic excitation of the defect can be the source of its motion. Such a process called by Stoneham the local excitation induced reaction [56,57] can occur if there is strong diminishing or even cancelation of the motional barrier at the excited state of the defect. If this occurs, defect motion becomes an athermal or quasiathermal process. Examples of local excitation type processes are: charge-state dependence of diffusion rates in Si [58] or breaking bonds at the amorphous solids [59,60].

LLR induced defect production is best documented and understood for F - center production by anihilation of excitons in alkali halides [24,25,61,63]. Recently Williams and coworkers unequivocally proved that such a process really occured [61,63]. As it has been mentioned at the introduction, a hole creation in ionic crystals is almost invertely accompanied by its self-trapping. A molecular like complex of the two anions attracted by a hole in a (110) direction is therefore created [64]. Similar effect occurs in an exciton formation. In most cases a free exciton practically momentarilly relaxes, with a self-trapped hole being a main source of this relaxation process (Fig. 15a,b). Such a state is a precursor of a further instability causing anihilation of relaxed exciton into a pair of defects - n.n. F centre and an H - center (interstitial relaxed anion [65]). This has been shown by Williams and coworkers using technique of the excited state spectroscopy [61,63]. After relaxation the exction is in a relatively long lived triplet state (Fig. 15b). Light absorption to a nearest triplet excited state causes immediate annihilation of the exciton with creation of a pair of F center and nearly H center (Fig. 15c). At a moment of a pulsed laser excitation, absorption signal of STE decays and the F-center absorption increases. It is important to note that a triplet excited state is not itself a source of a defect production. An additional barrier along (110) axis must be overcomed to produce a defect pair. It can be done either via nonradiative recombination after a second light pulse, at low temperatures, or just by increase of temperature which would help overcome the barrier in a normal activation process.

Production of defects by annihilation of excitons at temperatures above 200 K causes darkening of a crystal. It has been shown in a very simple experiment on KBr crystals [66]. These crystals possessed already some concentration of the point defects decorated by precipitation of Au metallic particles. After exciton annihilation these particles simply grew up.

Similar process but in covalent solids has been observed by Lang and Kimmerling (for a through review see Lang's paper [22]) [54,55]. In an irradiated GaAs one of the levels, being 0.31 deep electron trap,

DEFECT PRODUCTION BY EXCITONS
IN ALKALI HALIDES

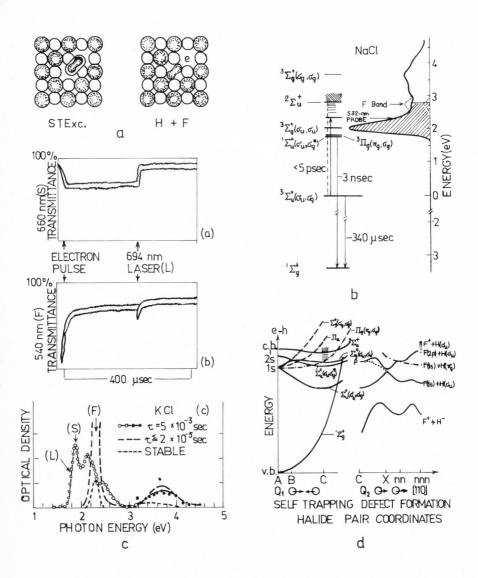

Fig. 15. Defect production by excitons in alkali halides. a) microscopic model of a self-trapped-exciton and n.n. H and F defect created after decay of a STE state; b) energy diagram of STE absorption measurement [54]; c) direct experimental proof of creation of F-centers (F) accompanied by STE annihilation (S) [52]; d) configuration coordinate model of (left) self-trapping of excitons and (right) defect formation in alkali halides [52,54].

can be thermally annealed. The activation energy for a purely thermal process is 1.4 eV (Fig. 16). When at such defects, already present in a junction area, nonradiative recombination of electrons (filling this level) with injected holes takes place, their concentration dramatically decreases. Now annealing activation energy is only 0.34 eV (Fig. 16). A difference in energy of 1.06 eV had to be supplied to a lattice to cause enhancement of annealing process by a nonradiative recombination. As is seen from the right diagram of Fig. 16, electron-hole recombination energy is 1.09 eV, just enough for a mentioned process. In the crystals with higher band gaps, recombination energy is even higher, therefore recombination enhanced processes, resembling well known Auger process for charge carriers, must be more probable [22].

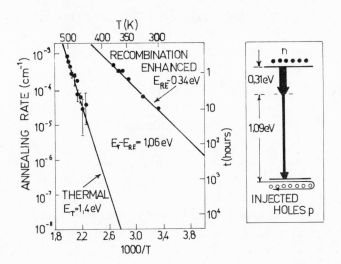

Fig. 16. Direct proof of recombination enhanced annealing of radiation-induced defects in GaAs. Additional energy is gained by nonradiative electron-hole recombination at defect during a forward bias [59].

Importance of this kind of solids tate reaction is obvious, since such processes could be a source of different degradation phenomena in solid state devices, especially at high injection levels. Moreover, defect production in ionic solids is of certain importance too, since it can be used e.g. in fabriaction of semipermanent optical memories.

SUMMARY

All the experiments presented above clearly indicate that the large lattice relaxation effects are in the same way operating in ionic as well as covalent crystals. It must be stressed, however that although the net results or effects (e.g. persistent conductivity) are the same, a detailed mechanism may be quite different in the both types of solids. One property of the defect must be preserved, however. This is a charge localization. It is now well established that the depth of the level in most cases bears no information about electron or hole localization. From all what has been presented in this review it seems to be well established that occurence of any type of lattice relaxation (defect motion, metastability effects, Stokes shift or just a tempera-ture broadening of the optical spectra) is a sign of such a charge localization at the defect. Necessity of a simultaneous incorporation of electronic as well as lattice interaction is quite challanging for theorists. Simple phenomenological models, as e.g. configuration coordinate model, are nothing more but a picture of the processes involved. Meaning of a configuration coordinate, not mentioning a dimensionality of the diagram are in most cases lacking. It is almost clear that most of the defects described above involve not simple sub-situtional impurities, but rather more complicated aggregates. A simplest model proposed by Lang et.al. [67,68] is just a pair of a chemical impurity donor and anionic vacancy. It is not obvious, however, that the reality is just so simple, although much evidence for cor-rectness of the model has been provided [67].

All these indicate necessity of more experimental efforts in revealing all the microscopic properties of the defects at which LLR occurs. The simplest in principle, but not easy in practice, is deter-mination of the symmetry of the defect before and after relaxation. EPR and ENDOR or both of these techniques coupled with the optical ex-citation and stress, could provide a desired information. Another very crucial point is a type of a lattice mode involved, as well as the values of force constants. Since in this class of defects only absorption measurement is possible, the only direct information available is about vibronic frequency of the defect in its filled state. Some information about the excited state can be given e.g. by a resonant Raman experiment, similar to this made on a conducting CdF_2:In [69]. After excitation the lattice is in a highly excited state, what manifests in a very struc-tured Raman spectrum (Fig. 17). Derivation of a vibronic frequency from this kind of experiment may be not very simple, because the final state

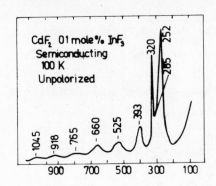

Fig. 17. Resonant Raman spectrum of semiconducting CdF_2:In (In^{2+} localized state
are present) [69].

is not stationary and because of possible strong anharmonicity at so highly excited lattice state (E_{relax} is in this case about 1.7 eV).

The final conclusion of this brief review is a statement that defects exhibiting strong coupling to the lattice, treated up to now as a s-ort of a scientific curiosity, at least in covalent crystals, are more common perhaps, than a simple guessing based on a number of experimental evidences at present would indicate.

REFERENCES

1. M.R.Lorenz, B.Segall and H.H.Woodbury, Phys.Rev. 134, A751 (1964)
2. M.C.Craford, G.E.Stillman, J.A.Rossi and N.Holonyak Jr., Phys. Rev. 168, 867 (1968)
3. H.C.Wright, R.J.Downey and J.R.Canning, J.Phys.D. 1, 1593 (1968)
4. R.J.Nelson, Appl.Phys.Lett. 31, 351 (1977)
5. D.V.Lang and R.A.Logan, Phys.Rev.Lett. 39, 635 (1977)
6. S.Porowski, M.Konczykowski and J.Chroboczek, phys.stat.sol.(b) 63, 291 (1974)
7. M.Baj, L.Dmowski, M.Konczykowski, S.Porowski, phys.stat.sol.(a) 33, 421 (1976)
8. A.Ya.Vul, L.V.Golubev, L.V.Sharonova and Yu.V.Shmartsev, Fiz. Tverdogo Tela 4, 2347 (1970)
9. U.Piekara, J.M.Langer and B.Krukowska-Fulde, Solid State Commun. 23, 583 (1977)
10. G.Vincent and D.Bois, Solid State Commun. 27, 431 (1978)
11. D.Bois, A.Chantre, G.Vincent and A.Nouailhat, Inst.Phys.Conf.Ser. No.43, 277 (1979)
12. A.Mittoneau and A.Mircea, Solid State Commun. 30, 157 (1979)
13. G.W.Iseler, J.A.Kafalas, A.J.Strauss, H.F.MacMilland and R.M.Bube, Solid State Commun. 10, 619 (1972)
14. M.Baj, Ph.D.Thesis, Warsaw University 1979, to be published
15. A.Ya.Vul and A.Ya.Shik, Fiz.Tverdogo Tela 8, 1952 (1974)
16. A.Ya.Shik and A.Ya.Vul, ibid. 8, 1675 (1974)
17. A.Ya.Vul, Sh.I.Nabiev, S.G.Petrosyan, L.V.Sharonova and A.Ya.Shik, Fiz.Tekch.Poluprov. 11, 914 (1977)
18. L.Dmowski, M.Baj, M.Kubalski, R.Piotrzkowski and S.Porowski, Inst. Phys.Conf.Ser. No.43, 417 (1979)
19. A.G.Milnes, "Deep Impurities in Semiconductors", Wiley (N.Y.)1973.
20. H.G.Grimmeis, Ann.Rev.Matl.Sci. 7, 431 (1977)
21. J.M.Langer, U.Ogonowska and A.Iller, Inst.Phys.Conf.Ser. No.43, 277 (1979).

22. D.V.Lang, Inst.Phys.Conf.Ser. No.31, 70 (1977); part 5 of this review gives up-to date summary of recombination-enhanced defect motion in semiconductors.
23. Y.Toyozawa, Progr.Theoret.Phys.Suppl. 12, 111 (1959)
24. Y.Toyozawa, Proc. 4th Int.Conf.Vac. UV Rad.Phys. Hamburg, p.317 (Pergamon and Vieweg) 1974.
25. Y.Toyozawa, Proc.Southampton Conf. on Recomb. in Semicond., 1978 to be published (Sol.St.Electron.)
26. D.Emin and T.Holstein, Phys.Rev.Lett. 36, 323 (1976)
27. W.Beal Fowler, ed. "Physics of Color Centers" (Academic Press, New York) 1968.
28. W.Beal Fowler, Phys.Rev. 135, A1725 (1964)
29. Y.Kondo and H.Kanazaki, Phys.Rev.Lett. 34, 664 (1975)
30. R.F.Wood and U.Öpik, Phys.Rev. 175, 783 (1969)
31. R.G.Gilbert and R.F.Wood, J.Lminescence 1,2, 619 (1970)
32. Y.Toyozawa, J.Luminescence 12/13, 13 (1976)
33. J.M.Langer, 1976 (unpublished)
34. M.Kukimoto, C.H.Henry and F.R.Merritt, Phys.Rev. B7, 2486, 2499 (1973)
35. C.H.Henry and D.V.Lang, Phys.Rev. 1315, 989 (1977)
36. C.W.Struck and W.H.Fonger, J.Chem.Phys. 64, 1784 (1970); and former papers cited therein
37. R.Pässler, phys.stat.sol.(b) 86, K39 (1978)
38. C.W.Struck and W.H.Fonger, J.Luminescence 1,2, 456 (1970)
39. F.S.Ham, Phys.Rev. 138,A1727 (1965)
40. U.Fano, Phys.Rev. 124, 1866 (1961)
41. J.C.Phillips in Proc.Int.Scholl.Phys. "Enrico Fermi", Course 34 (Academic Press, N.Y.) 1976.
42. Y.Toyozawa, M.Inoue, T.Inui, M.Okazaki and E.Hanamura, J.Phys. Soc.Japan 22, 1317, 1349 (1967)
43. J.M.Baranowski, J.M.Langer and S.Stefanova, Proc. XI.Int.Conf. Semicond. 2, 1001, ed.M.Miasek, PWN, Warsaw (1972)
44. A.Radliński, phys.stat.sol.(b) 84, 503 (1977)
45. M.Kamińska, J.M.Baranowski, M.Godlewski, Inst.Phys.Conf.Ser.No.43, 303 (1979)
46. M.Kamińska, Ph.D.Thesis, Warsaw University (1979)
47. U.Ogonowska, this conference
48. J.M.Langer, T.Langer, G.L.Pearson, B.Krukowska-Fulde and U.Piekara, phys.stat.sol.(b) 66, 537 (1974); Solid State Commun. 13, 767 (1973)
49. L.Dmowski, M.Konczykowski, R.Piotrzkowski and S.Porowski, phys. stat.sol.(b) K131 (1976)
50. M.Konczyakowski, S.Porowski and J.Chroboczek, High.Temp.-High Press. 6, 111 (1974)
51. B.C.Burkey, R.P.Khosla, J.R.Fisher and D.L.Losee, J.Appl.Phys. 47, 1095 (1976)
52. R.Legros, Y.Marfaing and R.Triboulet, J.Phys.Chem.Solids 39, 179 (1973)
53. J.D.Weeks, J.C.Tully and L.C.Kimmerling, Phys.Rev. B12, 3286 (1975)
54. D.V.Lang and L.C.Kimmerling, Phys.Rev.Lett. 33, 489 (1974)
55. L.C.Kimmerling, D.V.Lang, Inst.Phys.Conf.Ser. 16, (London) p.589 (1975)
56. A.M.Stoneham, "Theory of defects and defect processes" AERE preprint TP 759, October 1978 (to be published)
57. A.M.Stoneham, C.R.Catlow and P.W.Tasker, Solid State and Electr. Dev. 2, 583 (1978)
58. G.D.Watkins, in "Radiation Effects in Semiconductors", ed. F.Vook (Plenum, N.Y.) p.67, 1968.
59. R.A.Street, Solid State Commun. 24, 363 (1977)
60. R.A.Street, Inst.Phys.Conf. Ser. No.43, 1291 (1979)
61. R.T.Williams, Semicond. and Insulators 3, 251 (1978)

62. Ch.Lushchik, I.Vitol and M.Elango, Sov.Phys.Usp. $\underline{20}$, 489 (1977)
63. R.T.Williams, J.N.Bradford and W.L.Faust, Phys.Rev. $\underline{B18}$ (1978) and references cited therein
64. M.N.Kabler, in "Point Defects in Solids" ed. J.H.Crawford,Jr. and L.M.Slifkin, Vol.I. 327-380 (Plenum, N.Y.) 1972.
65. E.Sonder and W.A.Sibley, ibid. pp.201-290
66. H.V.Yigi, A.F.Malysheva, Ch.Lushchik, E.S.Teesler, Phys.Tverd. Tela $\underline{14}$, 117 (1976)
67. D.V.Lang, R.A.Logan and M.Jaros, Phys.Rev. $\underline{B19}$, 1015 (1979)
68. D.V.Lang, R.A.Logan, Inst.Phys.Conf. Ser. No.$\overline{43}$, 433 (1979)
69. M.P.O'Horo and W.B.White, Phys.Rev. $\underline{B7}$, 3748 (1973)

TEMPERATURE DEPENDENT DECAY OF A METASTABLE STATE OF SYSTEMS WITH LARGE IMPURITY-LATTICE RELAXATION (CdF$_2$: In)

U. Ogonowska
Institute of Physics,
Polish Academy of Sciences
02-668 Warsaw, Al.Lotników 32/46, Poland

INTRODUCTION

The defect-lattice interaction leads to important modification of the pure electron models of many phenomena in doped crystals.

Recently many authors adopted a "configuration" model of deep impurities phenomena as: photoionization of impurity centers [1,2], metastable effects [3,4] and trapping of the free carriers by deep centers [5,6].

Quantitative description of the last two effects requires a good model of a temperature dependent transition probability through the barrier separating the states in the configuration coordinate model.

The aim of this paper is to show the breakdown of the classical Mott's picture [7], in which transition probability through the barrier height U is given by a formula

$$P(T) = A \cdot \exp(-U/KT) \tag{1}$$

where A is a constant of order $10^{13} s^{-1}$, which is nothing more but classical thermal overcoming the barrier.

Calculations are performed using parameters characteristic for the CdF$_2$:In case, which is a well argumented example of the metastable state relaxation.

This paper is a continuation of our papers [8,9], in which the configuration model of indium impurity in CdF$_2$ crystals was proposed and analysed. The main advantage of the present approach is incorporation of the two characteristic vibronic frequencies for the ground and excited states of the system. In a previous paper [9] one effective frequency for both states was used.

CONFIGURATION MODEL OF INDIUM IMPURITY IN CdF$_2$ CRYSTALS

Due to a different screening of an In^{3+} core by localized In$^{2+}_{loc}$ or a delocalized In$^{2+}_{hydr}$ electron, the equilibrium lattice position around impurity must be different in the both states.

However, for the In$^{2+}_{hydr}$ state due to its delocalized wave function the lattice position is expected to be the same as for the ionized In^{3+} center. These simple arguments lead to the configuration coordinate (C.C.) model (Fig. 1) [8,9].

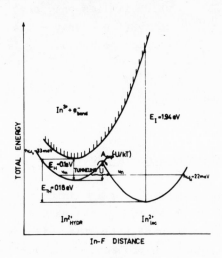

Fig. 1. Configuration diagram of CdF$_2$:In

Two absorption bands are observed in CdF$_2$:In crystals. The near infrared (IR) band is due to a hydrogenic state photoionization, the second in the visible and near ultraviolet region (UV) is caused by a photoionization of the localized state.

Theoretical interpretation of position and shape of these two bands and transport measurements provided a complete set of parameters of the C.C. diagram (E_H, E_I, E_{TH}, $\hbar\omega_u$), which had been used for theoretical calculation in this paper. The vibrational frequency for the In$^{3+}_{hydr}$ state is taken from the Raman effect measurement [10].

METASTABLE EFFECTS

At low temperatures (T < 50 K) the IR absorption band and electrical conductivity vanish due to electron freezing out on the deeper In$^{2+}_{loc}$ state.

Exposure to visible and near ultraviolet light produces non-equilibrium IMR absorption and conductivity which are very stable in low temperatures T < 50 K. In the 50 K < T < 110 K temperature region relaxation to the equilibrium state is observed with the time constant strongly decreasing with increase of temperature.

The relaxation kinetics is governed by a barrier U (Fig. 1). The experimental data taken in the temperature region 70 K < T < 110 K are shown on Fig. 2.

The activation energy of 1/T temperature dependence changes from

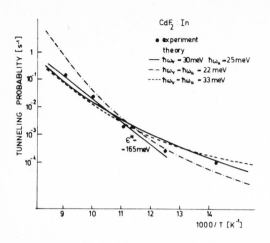

0.08 eV to 0.17 eV. Even the largest is almost twice smaller than the geometrical barrier height -0.3 eV calculated from the C.C. diagram parameters.

This phenomenon cannot be explained by the classical Mott's picture, which implies only one activation energy - the barrier height U.

Fig. 2.   Temperature dependent decay of a metastable In^{3+}_{hydr} state. Comparison between tunneling theory and experimental data is presented.

"TUNNELING" MODEL APPLICATION TO BARRIER CROSSING IN CdF_2:In

Recently Struck and Fonger (S.F.) have developed a quantum-mechanical model of nonradiative transitions between two states displaced horizontally in a configuration coordinate [11]. The validity of the Condon approximation with the transition probability proportional to the squared overlap integral of vibronic functions (Fig. 1) has been assumed:

$$N_{nm} = N_{uv}(1-r_v)r_v^m * <u_n v_m>^2 \qquad v_m \rightarrow u_n \qquad (2)$$

N_{nm} must be summed over all initial vibrational levels. The final levels are th se which conserve energy (transitions almost perfectly horizontal)

$$h\omega_o + mh\omega_v - nh\omega_u + h\omega_{latt} = 0$$

The nonradiative balance needs at least one phonon from the continuum of the lattice to secure the conservation of the energy. In the eq.(2).
N_{uv} - is the constant from the electronic partion of the transition integral, of the order $10^{13}s^{-1}$ near the constant in Mott's formula (1)
$h\omega_v$, $h\omega_u$ - vibrational frequences in initial and final states

$r_v = \exp(-h\omega_v/kT)$ - the Boltzmann factor between initial vibrational
levels.

$h\omega_o$ - is the energy of the v - parabola $m = 0$ level above the
u - parabola $n = 0$ level.

Overlaps $\langle u_n | v_m \rangle$ depend on the two parameters: the parabola's offset
and the ratio of the parabola's force constants $h\omega_v/h\omega_u$. For the
equal force constants the formula (2) simplifies and we do not need to
calculate overlaps explicitly [11]. For the $h\omega_v \neq h\omega_u$ case we have
to calculate probability from the (2) formula and to sum over all
initial states. Succesive overlaps are calculated numerically, with
the aid of recursion formulae given by Struck and Fonger [11].
The principal difference between S.F. quantum model and classical
Mott's model is that the S.F. properly takes into account tunneling
processes of the barrier crossing. Tunneling is possible because of
the wave-function overlap. Since population of the tunneling levels
strongly depends on temperature an effective barrier height must depend
on the temperature too. The temperature dependence of the activation
energy (ϵ^*) is clearly seen on Fig. 3. U^2_{pum} and r_v^m versus initial
state number are showed (U^2_{pum} is
defined in [11]), for the
$h\omega_v = 0.030$ eV, $h\omega_u = 0.025$ eV.
$U^2_{pum} * r_v^m$ product as a function
of m has a maximum which goes to
the higher number m when the temperature rises. We can notice that
for T = 100 K the main contribution to the tunneling comes from
the m = 5 level what corresponds
well with the experimental activation energy in that temperature
$\epsilon^* = 0.165$ eV.

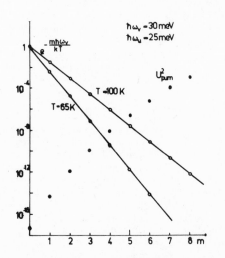

Fig. 3. U^2_{pum} and the Boltzmann factor
$e^{-mh\omega_v/kT}$ versus the initial
vibrational state number.

Due to relative simplicity
and analytical form the S.F. model
is suitable in fits to the experimental data. It should be
stressed that at the high tempera-
ture limit the S.F. model approaches the classical Mott's rate.

The model was used originally to interpret temperature quenching
of luminescence Tm^{3+} in La_2O_2S [11], and agreement between experimental data and theory was very good.

Transition through the barrier separating two states on C.C.

diagram were studied by Henry and Lang [5]. They studied thermally
activated carrier capture by deep centers in GaP. The temperature dependence of capture cross section has in high temperature limit the
single exponential form, but at the sufficiently low temperatures this
dependence becomes slower. According to the quantum S.F. model this
flattening is caused by the tunneling transitions from the lower
initial vibrational level through the barrier.

REFERENCES

1. B.Monemar, L.Samuelson, Phys.Rev. B18, No 2, p.809 (1978) and
   references cited therein
2. M.Kamińska, J.M.Baranowski, M.Godlewski, Proc. 14th Int. Conf. on
   the Physics of Semicond. Edinburgh (1978)
3. S.Porowski, M.Kończykowski, J.Chroboczek, phys.stat.sol.(b) 63,
   291 (1974)
4. G.Vincent, D.Bois, Solid State Commun. 27, 431 (1978)
5. C.H.Henry, D.V.Lang, Phys.Rev. 15, 989 (1977)
6. D.V.Lang, R.A.Logan, Phys.Rev.Lett. 39, No 10, 635 (1977)
7. N.F.Mott, Proc.Roy.Soc. (London) A167, 384 (1938)
8. U.Piekara, J.M.Langer, B.Krukowska-Fulde, Solid State Commun. 23,
   583 (1977)
9. J.M.Langer, U.Ogonowska, A.Iller, Proc. 14th Int.Conf. onf the
   Physics of Semicond. Edinburgh (1978)
10. M.P.O'Horo, W.B.White, Phys.Rev. B7, No 8, 3748 (1973)
11. C.W.Struck, W.M.Fonger, J.Lumin. 10, 1 (1975)

ELECTRON-PHONON INTERACTION: POLARON TRANSPORT

J.T. Devreese
International Center for Theoretical Solid State Physics
Universitaire Instelling Antwerpen Department Natuurkunde

Universiteitsplein 1., B-2610 Wilrijk, Belgium

1. THE FRÖHLICH POLARON HAMILTONIAN AND THE SPECTRUM

By now the Fröhlich Hamiltonian is well-known by most solid state physicists and field theorists:

$$\hat{H} = \frac{\hat{p}^2}{2m} + \sum_k \hbar\omega_o \hat{a}_k^+ \hat{a}_k + \sum_k (V_k \hat{a}_k e^{i\bar{k}.\hat{r}} + V_k^* \hat{a}_k^+ e^{-i\bar{k}.\hat{r}}) \tag{1a}$$

$\hat{r}$ and $\hat{p}$ are the electron position, respectively momentum operators, $\hat{a}_k^+$ and $\hat{a}_k$ are the creation and annihilation operators for longitudinal optical phonons with wave vector $\bar{k}$ and energy $\hbar\omega_o$ while V_k is the matrix element for the electron-phonon coupling:

$$V_k = i \frac{\hbar\omega}{k} \sqrt{\sqrt{\frac{\hbar}{2m\,\omega_o}}} \sqrt{\frac{4\pi\alpha}{V}} \tag{1b}$$

V is the volume of the crystal and α is the Fröhlich polaron coupling constant

$$\alpha = \frac{e^2}{\hbar c} \sqrt{\frac{mc^2}{2\hbar\omega_o}} \left(\frac{1}{\varepsilon_\infty} - \frac{1}{\varepsilon_o}\right) \tag{1c}$$

As yet nobody has succeeded to diagonalize the Hamiltonian (1a) exactly.

Standard approximation results for the binding energy and the effective mass of the polaron are displayed in Table I.

1.1 The optical spectrum of Fröhlich polarons

In Ref. [1] we have calculated the optical conductivity Re $\sigma(\Omega)$ of a polaron using the Feynman model [2]. The optical conductivity reflects the spectrum of the polaron Hamiltonian (1a). In Fig. 1 this optical density is plotted for $\alpha = 6$. Three kinds of excitations are seen to appear in polarons:
1) the creation of a longitudinal optical phonon in the crystal by the incident light (denoted by "PS", phonon scattering in Fig. 1).
2) relaxed excited states ("R.E.S."); these are internal excited states for which the lattice is readapted (relaxed) to the distribution of the excited electrons.

Table I. Calculation methods of the polaron self energy and effective mass

---

Perturbation theory		

Second order [20]	Weak coupling	$\frac{\Delta E}{\hbar\omega} = -\alpha$ $\qquad$ $\frac{m^{\bigstar}}{m} = (1-\frac{\alpha}{6})^{-1}$

Canonical transformations		

Elimination of the electron coordinates [21]	Weak coupling	$\frac{\Delta E}{\hbar\omega} \leqslant -\alpha$ $\qquad$ $\frac{m^{\bigstar}}{m} = 1 + \frac{\alpha}{6}$
Ibidem and diagonalization of recoil [22]	Weak coupling	$\frac{\Delta E}{\hbar\omega} \leqslant -\alpha -1.26(\frac{\alpha}{10})^2$ $\quad$ $\frac{m^{\bigstar}}{m} = 1+\frac{\alpha^2}{6}+2.24(\frac{\alpha}{10})^2$ $\quad -1.875(\frac{\alpha}{10})^3$
	Strong coupling	$\frac{\Delta E}{\hbar\omega} \leqslant -1.05\,\alpha^2$
Elimination of the center of gravity [23]	Strong coupling	$\frac{\Delta E}{\hbar\omega} \leqslant -\frac{\alpha^2}{3} - 3\log 2 - \frac{3}{4}$

Adiabatic approximation		

Localized trial wave functions [24]	Strong coupling	$\frac{\Delta E}{\hbar\omega} = -0.106\,\alpha^2$ $\quad$ $\frac{m^*}{m} = 0.0208\,\alpha^4$
Translational invariant trial wave function [25]	Weak coupling	same results as [21]
	Strong coupling	$\frac{\Delta E}{\hbar\omega} = -0.106\,\alpha^2$ $\quad$ $\frac{m^*}{m} = 155(\frac{\alpha}{10})^4(1+\frac{67}{\alpha^2})^{-1} -2.73-\frac{8.6}{\alpha^2}$

Green's functions		

Summation of diagrams [26]	Weak coupling	same results as [21]
Variational solution of the Dyson equation [27]	Weak coupling	same results as [21]
	Strong coupling	$\frac{\Delta E}{\hbar\omega} = -0.108\alpha^2 - 3\,\ln 2 - 0\,(\alpha)^2$

Path integrals		

| Elimination of phonon variables and quadratic approximation [2] | All coupling | $\frac{\Delta E}{\hbar\omega} \leqslant \frac{3}{4v}(v-w)^2$ $-\frac{\alpha v}{\sqrt{\pi}} \int_0^\infty \frac{du\ e^{-u}}{[w^2 u + v^{-1}(v^2-w^2)(1-e^{-uv})]^{1/2}}$ $\frac{m^{\bigstar}}{m} = 1 + \frac{1}{3}\,\pi^{1/2}\,\alpha\,x$ $x \int_0^\infty \frac{du\ u^2 e^u}{[w^2 u + v^{-1}(v^2-w^2)(2-e^{-uv})]^{3/2}}$ v and w are variational parameters |

Table II. (continued)

Heisenberg equation of motion

Perturbative solution [28]	Weak coupling same result as [20]
Self-consistent solution [9]	Weak and strong coupling

$\dfrac{\Delta E}{\hbar \omega}$: same result as [21] and [24] respectively

$$\frac{m^*}{m} = \left[1 - \frac{\alpha}{3\sqrt{\pi}} \left(\frac{m}{m}\right)^{1/2} \int_0^\infty du \; \frac{u^2 \; e^{-u}}{[C(1-e^{\xi u})+u]^{3/2}} \right]^{-1}$$

C and ξ are determined self-consistently

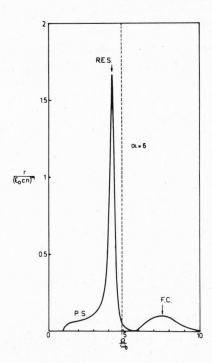

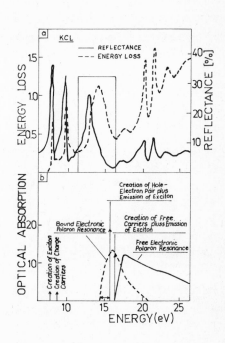

Fig. 1. The optical absorption of polarons for α=6; at T=0. The first
peak from the l.h.s. is a pho-
non-scattering state. The second
peak from the l.h.s. is the relaxed excited state. The third
peak from the l.h.s. is the so-
-called Franck-Condon state.

Fig. 2. a) Experimental reflectance and
energy loss for KCl (from D.M.
Rossler and W.C.Walker, Phys.Rev.
160, 599 (1968))
b) The calculated optical absorption using the electronic
polaron concept.

3) Franck-Condon excitations (F.C.) for which the electron is excited
   without the occurence of lattice relaxation. For weak coupling the
   phonon scattering peak is dominant while for larger coupling the
   R.E.S. takes most of the oscillator strength. The transition between both regimes is rather abrupt at $\alpha \simeq 5.8$.

It is remarkable that for sufficiently high coupling almost all
the oscillator strength for optical absorption by polarons goes into
one continuum state: the R.E.S. Our recent calculations of self-consistent improvements to the optical spectra of polarons (as proposed
by Thornber [3]) indicate that these are very small for all α. [4].

1.2 Electronic polaron

The "electronic polaron", which is an electron surrounded by the
polarization field of longitudinal excitons, is useful in describing
the far ultraviolet and the soft X-ray absorption, e.g. in alkali
halides. In [5] we have applied our results for the optical absorption
of Fröhlich polarons to these spectra (Fig. 2).

2. OHMIC TRANSPORT OF POLARONS

Also the calculation of the mobility of polarons in the Ohmic
limit is a standard problem. In Table II. we review some results. Despite the efforts which have been devoted to the polaron mobility
problem, difficulties remain. Such a difficulty is that the theories are
based on the relaxation time approximations differ from those obtained by
Feynman, Hellwarth, Iddings, Platzman [6] and Thornber & Feynman [7].

The calculation of Feynman et.al. (FHIP) [6] is not based on the
relaxation time approximation. This time-dependent calculation is
based on the Kubo formula adapted to path integrals. The polaron is
described by the Feynman model where the phonon field is simulated by
a fictitious particle. The difficulty with FHIP seems to occur when
the limits $\nu \to o$ (ν is the frequency of the incident radiation) and
$\varepsilon \to 0$ (ε is the infinitesimal from the adiabatic approximation) are
taken. In FHIP the $\varepsilon \to o$ limit is taken first, before the $\nu \to o$ limit.

Also the relaxation time apprximation has its limitations. However, for low electric fields and weak coupling it is expected to be
valid [8]. This conclusion is corroborated by our recent finding [9]
that the rigorous solution of the Boltzmann equation leads to the
relaxation time mobility for small α .

<u>Table II.</u> Calculation methods of the polaron mobility

Relaxation time: $\mu = \dfrac{e\tau}{m^*}$; $\dfrac{1}{\tau}$ total transition probability per unit time

---

F.Low-D.Pines:

The Lee-Low-Pines wavefunctions are used to calculate the relaxation time τ.

$$\mu = \frac{e}{m}\frac{1}{2\alpha\omega_{LO}}\exp(\frac{\omega_{LO}}{kT})\frac{1}{(1+\frac{\alpha}{6})^3}\,f(\alpha) \qquad (\hbar = 1)$$

$f(\alpha)$ is a parameter in the theory depending on the coupling constant (see Ref. [29]).

T.D.Schultz:

The relaxation time τ is calculated with the "golden rule". A Hamiltonian, equivalent with the Feynman trial action is used to determine the wavefunctions and energy spectrum:

$$\mu = \frac{e}{2m\omega_{LO}}\frac{1}{\alpha}(\exp(\frac{\omega_{LO}}{kT})-1)\frac{1}{z_r}\frac{m_o}{m}\frac{v_r}{(2\omega_{LO}/m)^{1/2}} \qquad (\hbar = 1)$$

z_r, v_r, m_o are parameters depending on the coupling constant (see Ref. [30])

D.Langreth:

This is a modification of Low and Pines, the relaxation time is calculated relying on the optical theorem, the wave functions are calculated with perturbation theory

$$\mu = \frac{e}{2\alpha\omega_{LO}m}(\exp(\frac{\omega_{LO}}{kT})-1)\,f(\alpha)\,(\frac{m}{m^*}) \qquad (\hbar = 1)$$

$f(\alpha)$, m depend on the coupling constant [31]

E.Kartheuser, J.Devreese, R.Evrard:

The total transition probability is calculated relying on the Chew-Low equations. These equations are solved with the self-consistent equation of motion method

$$\mu = \frac{e}{2\alpha\omega_{LO}m}(\exp(\frac{\omega_{LO}}{kT})-1)(\frac{m}{m^*})^{3/2}e^{\frac{m^*}{m}F(\alpha)}\,G(\alpha) \qquad (\hbar = 1)$$

$F(\alpha)$ and $G(\alpha)$ are given in Ref. [10]

Kubo formalism

Y.Osaka:

The current-current correlation function is obtained from a double path integral. The time evolution in this path integral is simulated by Feynman's dual action. Asymptotic expressions for the large time behaviour are used.

$$\mu = \frac{e}{2\alpha\omega_{LO}m}(\exp(\frac{\omega_{LO}}{kT})-1)\exp(m_o\eta/\nu)/(m_o)^{3/2}+0(kT) \qquad (\hbar = 1)$$

m_o, η, ν are parameters depending on α and kT. [32]

Table II. (continued)

A.Weyland:
    The current-current correlation function is calculated using the perturbative approach of Van Hove (after the Lee-Low-Pines transformations are performed).

$$\mu = \frac{e}{2m\alpha\omega_{LO}}(e^{\frac{\omega_{LO}}{kT}} -1)\,(1+C_1\alpha +C_2\alpha^2 + \ldots) \qquad (\hbar = 1)$$

    The constants C_1 and C_2 can be found in Ref. [33].

D.Langreth-
P.Kadanoff:
    The current-current correlation function, and the corresponding two particle Green's function is calculated using a diamagnetic perturbative expansion.

$$\mu = \frac{e}{2\alpha\omega_{LO}m}\,(e^{\frac{\omega_{LO}}{kT}} -1)\,(1-\frac{\alpha}{6}) \qquad \hbar = 1\,(\alpha \rightarrow o) \qquad [34]$$

        Boltzmann equation

D.Langreth:
    The Kadanoff-Baym equation.

    The T-matrix, which determines the transition probabilities per unit time, has been calculated using a perturbative treatment for the so-called "self-energy" term of this matrix.

$$\mu = \frac{e}{2\alpha m^{\star}\omega_{LO}}\,(\exp\frac{\omega_{LO}}{kT} -1)\,(1+1.33\,\frac{kT}{\omega_{LO}}) \qquad \hbar = 1$$

$$m^{\star} = m\,\frac{1-0.0008\,\alpha^2}{1-\frac{\alpha}{6} + 0.003\mu\alpha^2} \qquad [7]$$

J.T.Devreese, R.Evrard, E.Kartheuser:

    The Boltzmann equation has been used, the mobility is calculated from $\vec{j} = \int d^3p\,\frac{e\vec{p}}{m}\,f(\vec{p})$.

    The transition probabilities in the Boltzmann equation are obtained from perturbation theory. The Boltzmann equation is solved with a method developed by J.T.Devreese and R.Evrard (Ref. [17]).

    The electric field dependence of the mobility is calculated numerically. For a weak electric field the results are close to the mobility found by the standard relaxation time approximation [8].

Table II. (continued)

Path integral methods

Feynman, Hellwarth, Iddings, Platzman:

The impedance $z(\omega)$ is calculated directly.

$z(\omega)$ has been derived from a double path integral. The time evolution was approximated by the Feynman trial action. For $z(\omega)$ a "self-energy" type correction has been made.

$$\mu^{-1} = \frac{2}{3} \frac{\alpha}{\sqrt{\pi}} \frac{\beta}{\sinh(\beta/2)} \left(\frac{v}{w}\right)^3 \frac{1}{2\sqrt{a^2-b}} K_1(\sqrt{a^2-b})$$

K_1 is the modified Bessel function of order 1. a,b,v,w are parameters given in Ref. [6] ($\hbar = m = \omega_{LO} = 1 = e$)

Thornber-Feynman:

$$\vec{E} - <\vec{p}> = \sum_k \vec{k} <R_k>$$

$\vec{k}R_k$ expresses the net rate of change of electron momentum, due to emission and absorption of phonons.

A steady state regim $<\vec{p}> = o$ is imposed. The path integral formalism is used to express $<R_k>$ as a function of the steady state velocity.

$$\frac{1}{\mu} = \frac{2}{3} \left(\frac{1}{A^{3/2}}\right) \frac{\beta}{\sinh(1/2\,\beta)} \frac{1}{\sqrt{C}} \left(\frac{\beta}{r}\right)^{1/2} K_1\left(\frac{1}{2}\beta\sqrt{C}\right) \quad m = \hbar = \omega_{LO} = 1$$

K_1 is the modified Bessel function of order 1 . The parameters A and C are functions of the coupling constant and the temperature [12].

It is interesting to note that if a drifted Maxwellian distribution for the electron momenta is imposed together with momentum conservation in the Boltzmann equation then the same mobility is obtained as that found by FHIP.

Recently we have used a self consistent equation of motion approximation (S.C.E.M.A.) [10] to describe the polaron. This self-consistent problem leads to a description which is similar to the Feynman model (especially for $\alpha \lesssim 4$ and $\alpha \to \infty$). The operators describing the polaron obtained in this way can easily be used to perform a relaxation time calculation of the polaron mobility [11] . Some details of the derivation are presented in the Appendix. The main results of this calculation are

$$\mu_{sc} = \mu_0 \left(\frac{m}{m^*}\right)^{3/2} \exp\left[\frac{m^*}{m} F(\alpha)\right] G(\alpha) \tag{2a}$$

m^* is the polaron mass, $\bar{N} = \dfrac{1}{e^{\hbar\omega/kT}-1}$, $\mu_o = \dfrac{1}{2m\alpha\omega_o\bar{N}}$

$$G(\alpha) = \left[I_0\left\{\frac{m^*}{m}F(\alpha)\right\} + \sum_{n=1}^{\infty}\left[\exp\left(- \frac{n\Omega_o}{\omega}\frac{m^*}{m}F(\alpha)\right)/\left((1+n)\frac{\Omega_o}{\omega}\right)^{1/2}\right] I_n\left\{\frac{m^*}{m}\Gamma(1+\frac{n\Omega_o}{\omega})F(\alpha)\right\}^{-1}\right]$$

(2b)

$$F(\alpha) = \frac{2}{3}\frac{m\,\omega_o}{\hbar}\int d^3k\left(\frac{k}{m}\right)^2 |f_k|^2 \gamma_k^{-2}$$

(2c)

For weak coupling μ_{sc} tends to the well known result

$$\mu = \mu_o\left[1 - \frac{\alpha}{6} + 0(\alpha^2)\right]$$

(3)

The significance of the results, eqs(2)-(3) is that they extend the Kadanoff results [12] for intermediate coupling (eqs.(2) can be used for $\alpha \lesssim 4$).

In Table III. the different mobility calculations are compared with experiment. It is seen that the different theories lead to quite different results. The case of KCl is not completely understood at present. In analyzing Table III. it should be kept in mind that a rigorous mobility calculation can never produce results lower than experiment because not all scattering mechanisms are included in the theory.

Table III. Comparison of experimental results of polaron mobilities with that of different theoretical calculations in the case of silver and potassium halides. T is the temperature at which the experiment has been performed.

μ (cm^2/V.sec)	AgBr [a]	AgCl [b]	KBr [c]	KCl [d]
FHIP [6]	295	151	59	79
FT [12]	175	103	49	78
LPM [29,31]	190	188	73	261
Langreth [7]	383	294	98	239
K.D.E. [10]	184	181	63	217
$\theta = \dfrac{\hbar\omega_{LO}}{k}$ in $^\circ K$	200	283	243	310
T in $^\circ K$	164	152	154	97
Experiment [a,b,c,d]	142	130	51	102

a) reference [35]
b) reference [36]
c) reference [37]
d) reference [38]

3. NON-OHMIC POLARON TRANSPORT

To treat polaron transport in the non-Ohmic regime at all electron-phonon coupling strengths one needs to study the Liouville equation for the density matrix. Only in the limit of weak coupling and not too high fields the Boltzmann equation can be used.

Thornber and Feynman [7] have devised an approximate method to solve the Liouville equation for polarons. This method again is based on the path integral formulation of quantum mechanics which allows for a straightforward elimination of the phonon variables from the equation of motion. Assuming that a steady state is reached (and therefore not including the possibility of Fröhlich run-away solutions) the balance equation becomes

$$\vec{E} = \sum_{\vec{k}} \vec{k} < R_k >$$
(4)

where $\vec{E}$ is the applied electric field and $< R_k >$ is a quantum statistical average which measures the next rate of emission of L.O. phonons of wave vector $\bar{k}$ by the polaron. $< R_k >$ takes the form of a double path integral in the T.F. model. $< R_k >$ is then calculated for the Feynman polaron model (quadratic action).

Eq.(4) is conceptually very transparent and for weak coupling it is easily obtained from detailed balance and the golden rule. $< R_k >$ does not depend on $\vec{E}$ but only on α; its evaluation then proceeds in an analogous way to that of the path integrals occurring in [6]; the magnitude of $|\vec{E}|$ barely influences the complexicity of the calculation and this is a major advantage of the elegant expression (4) which, it may be restated, is valid at all α. In Fig. 3 the reported comparison between experiment and T.F. theory is shown for KBr [13] and SiO_2 [14] in the non-Ohmic field region.

The TF theory is the only known non-Ohmic polaron theory for intermediate and large α. Despite its successes the TF approximation has some drawbacks. In the limit E → o it reduces to the FHIP approximation and the well known difficulties regarding the factor $\frac{3}{2}$ kT arise [15]. "Faute de combattants" no theoretical test for the TF method is as yet available in the non-Ohmic regime for relatively large α.

For small α and not too high electric fields however the Boltzmann equation is valid. It is then possible to obtain some accurate results in this limit and compare them both with the experiment and with the TF weak coupling limit.

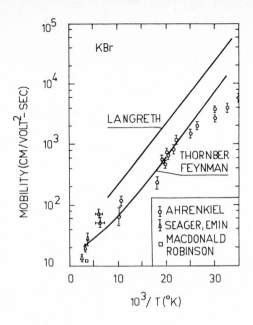

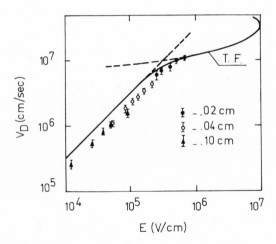

Fig. 3. a) The measured mobility of KBr is compared with the calculated mobility (from Ref. [13]).

b) The measured drift-velocity in the non-linear regime compared with the theoretical curve of Ref. [12] (from Ref. [14]).

Weak coupling and the Boltzmann equation

The Boltzmann equation for the electron momentum distribution can be written in the form

$$\frac{\partial f(\bar{p})}{\partial p_z}\, eE = \int f(\bar{p})\pi(\bar{p}\to\bar{p})\, d^3 p' - \int f(\bar{p})\pi(\bar{p}\to\bar{p})\, d^3 p' \qquad (4)$$

$\pi(\bar{p}\to\bar{p}')$ is the probability for a transition from $\bar{p}$ to $\bar{p}'$. It should be emphasized that (4), which already represents a formidable mathematical problem, is only valid for weak coupling. However this Boltzmann equation is extremely useful for the practical study of polar semiconductors because for most of them $\alpha \ll 1$.

Of course several *numerical* techniques (see e.g. [16]) have been developed to study the Boltzmann equation (4): e.g. Monte Carlo methods and iterative schemes. For weak electric fields the Monte Carlo method is not accurate because of the fluctuations in the electron motion. For relatively high electric fields both the Monte Carlo method and the iterative method are reliable and, as is well-known, have been applied successfully.

Why then is there a need for analytical methods to study the Boltzmann equation? First of all analytical results obviously contain more concise information in closed form than numerical methods. Furthermore analytical results concerning integral equations are interesting for their own sake. From the practical point of view the analytical study of the Boltzmann equation might be useful for the detailed study of avalanche. Indeed one then needs to know $f(\bar{p})$ for large $|\bar{p}|$ (like $\frac{p^2}{2m} > 10\ \hbar\omega_o$ in InSb). Manifestly Monte Carlo methods will fail to provide accurate values for $f(p)$ in regions where $f(p)$ is smaller than 10^{-5} times its maximum value as it is the case in the tails of the distribution which are, nevertheless, very important for avalanche. As far as we know, also the iterative method has not provided these accurate values for $f(p)$ in the "avalanche tail".

In Ref. [17] the present author and R.Evrard have devised a method to solve the time-independent Boltzmann equation analytically under the physical circumstance that $f(\frac{p^2}{2m} > 2\hbar\omega)=0$ (region III in Fig. 4). In many transport measurements this assumption is satisfied to a high degree of accuracy and in practice $f(\frac{p^2}{2m} > 2\hbar\omega)=0$ is no approximation at all.

Under this assumption the Boltzmann equation can be solved quasi-analytically at least for T=o. (Inclusion of T≠o can be treated starting from the T=o solution.) For T=o only *emission* of L.O. phonons

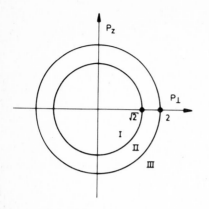

Fig. 4. The $(p_\perp, p_z)$ plane is divided in three regions.

can occur (due to the absence of real phonons) and

$$\pi^e (\bar{p} \to \bar{p}') = \frac{\alpha}{\pi\sqrt{2}} \frac{\delta\left\{\frac{p^2}{2}-1-\frac{p'^2}{2}\right\}}{|\vec{p}-\vec{p}'|^2} \quad (5)$$

The key to the solution of the Boltzmann equation now is the following [17]:

i) In the region II the first term of the r.h.s. of (4) is identically zero. This is because emission always brings electrons from II to I. Therefore if $p > \sqrt{2}$ one has $\pi (\bar{p}' \to \bar{p}) = 0$. In region II the Boltzmann equation therefore is

$$\frac{\partial f(p_\perp, p_z)}{\partial p_z} eE = \int_{p<\sqrt{2}} f(p'_\perp, p'_z) \pi (\bar{p} \to \bar{p}') d^3 p' \quad (6a)$$

ii) In the region I, the second term in the r.h.s. of eq.(4) is missing. This is because $\pi (\bar{p} \to \bar{p}') = 0$ if $p < \sqrt{2}$ as the electron does not have sufficient energy to emit a phonon and the Boltzmann equation there takes the form:

$$\frac{\partial f(p_\perp, p_z)}{\partial p_z} eE = -f(p_\perp, p_z) \int \pi (\bar{p} \to \bar{p}') d^3 p' \qquad p > \sqrt{2} \quad (6b)$$

At first sight eq.(6a) is still an integro-differential equation; however $f(p'_\perp, p'_z)$ is the distribution for $p' > \sqrt{2}$ and therefore it is the solution of eq.(6b). The problem is then solved as follows: first solve (6b) which is a first order differential equation and then introduce the solution into (6a) and integrate it to get the complete solution.

It should be noted that the above method is valid for T=0. It should be relatively easy to solve (6a) and (6b) rigorously. If it is assumed that $f(\bar{p})$ goes rapidly to zero when $p > \sqrt{2}$ some transparent closed forms can be obtained [17].

$$f(p_\perp, p_z) = f_0(p_\perp, p_z) e^{-g(p)} \qquad p > \sqrt{2} \quad (7a)$$

$$g(p) = \frac{2\alpha}{eE}\left\{p \cos^{-1}(\frac{p}{\sqrt{2}}) - \sqrt{p^2-2}\right\} \quad (7b)$$

and

$$f(p_1, p_z) = \int_{p_z^o}^{p_z} e^{-g(\sqrt{p^2+2})} \sqrt{p^2+2} \; dp_z \qquad p > \sqrt{2} \qquad (7c)$$

Continuity on the circle $p = \sqrt{2}$ is then achieved by taking $f(p_1, \sqrt{2-p_1^2})$ from (7c) as initial values for $f_o(p_1, p_z)$ in (7a). In Fig. 5 $f(p_1, p_z)$ is shown for InSb and for E = 17.8 V/cm.

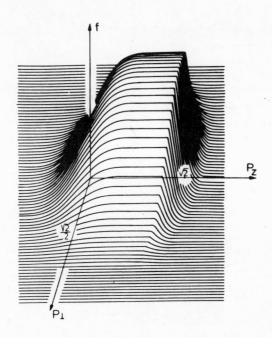

Fig. 5.   The momentum distribution function $f(p_1, p_z)$ for polarons in an external electric field in the p_z direction
at T=o, E=17.8 V/cm and $\alpha = 0.02$.

The results show that for $E \rightarrow o$ an essential singularity occurs in the solution for $f(\bar{p})$.

As will follow from the next paragraphs, the present method allows for practical applications.

In Ref. [8] we have proposed two iterative schemes one of which uses the results for $f(\bar{p})$ of the preceeding paragraphs (eq.(7)) as input. These iterative schemes then allow to study $f(p)$ at arbitrary temperature.

In Fig. 6 a comparison is made between theory and experiment for mobility of electrons in n-type InSb at 77°K. The full curve corresponds to our solution of the Boltzmann equation. For E → o the full line tends to $\mu = \dfrac{e}{2m\alpha\, N\omega}$ the standard result for Ohmic polaron conduction.

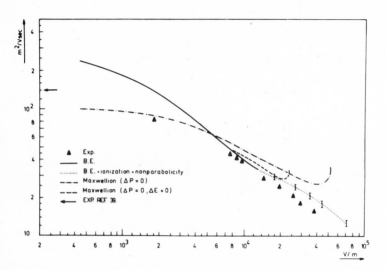

Fig. 6. Mobility of conduction electrons in n-type InSb at 77 K. ▲ Hall effect experiment, —— rigorous solution of the Boltzmann equation for parabolic band electrons (LO phonon scattering only), ... solution of the Boltzmann equation with ionization and non-parabolicity effects, -.- drifted Maxwellian for parabolic electrons (LO phonon scattering only) (Δ P=o), --- drifted Maxwellian for parabolic band electrons (LO phonon scattering only) (Δ P=o, Δ E=o), ← experiment (E → o) [39].

The effects of non-parabolicity and ionization of electron-hole pairs have been taken into account by introducing a "cut-off" at the critical momentum for ionization. For $E > 10^4$ V/m the agreement between theory and Alberga's recent experiments is rather satisfactory [18]. At lower electric fields the effects of impurity scattering and deformation potential scattering should be taken into account.

Although further calculation of the detailed ionization mechanism is required the following suggestions regarding the transport mechanism of polarons in InSb can be made:

1) Ionization plays an important role in transferring back electrons to low momenta so that they undergo scattering with LO phonons again and again. Therefore it can be stated that avalanche produces no dramatic kink in the log μ versus log E curve because of its efficiency in keeping electrons under the influence of the LO phonons.

2) Between 7×10^3 and 1.4×10^4 V/m electron-LO phonon scattering is the
   only scattering mechanism which has to be taken into account. This
   follows from the close agreement between experiment and theory
   (LO phonon scattering of parabolic band electrons) in this region.

Following Fröhlichs and Paranjape's ideas [19] one can express
the conservation of momentum in the polaron scattering processes and
choose a displaced Maxwellian for $f(\bar{p})$. The result is also shown in
Fig. 6. As indicated also by the arrow in Fig. 6 recent experimental
data at low field yield mobilities larger than the values predicted
for a drifted Maxwellian. This shows that, in this case, a drifted
Maxwellian is not a precise approximation in the Ohmic regime.

It may be noted that the Thornber-Feynman expression for the mobility in the limit of small α is equivalent to the mobility obtained
with a drifted Maxwellian.

CONCLUSION

The effective mass, self energy and optical absorption of Fröhlich
polarons can be calculated quite accurately at present. The Ohmic mobility at weak and intermediate coupling presumably is calculated most
accurately by the relaxation time method. For the non-Ohmic regime the
Boltzmann equation, for which analytical results are presented in this
lecture, is a practical tool while the Thornber-Feynman method remains
unchallenged for non-Ohmic behaviour if α is relatively large.

It remains necessary to elucidate the problem related to the
discrepant factor $\frac{3}{2}$ kT of the path integral polaron mobility calculations. Ref. [11] is an attempt in this direction because it combines
the relaxation time approximation with an intermediate coupling description.

APPENDIX: GENERAL FORMALISM FOR POLARON RELAXATION TIME (Ref. [10])

In the formal scattering theory one usually expresses the scattering cross section in terms of the transition matrix R. The total probability of the transition per unit time, i.e., the inverse of the relaxation time τ^{-1}, is then obtained from the cross section by integrating over all the possible initial and final states. For the polaron problem, this gives

$$\tau^{-1} = \frac{2\pi}{\hbar} \bar{N} \int d\vec{k}_o \int d\vec{k}_f \int d\vec{P}_f^* |< \vec{P}_f^*,\vec{k}_f| \; R \; |\vec{P}_o^*,\vec{k}_o >|^2$$
$$\times \; \delta \; [E(\vec{P}_f^*) - E(\vec{P}_o^*)] \tag{A.1}$$

where

$$\bar{N} = (e^{\theta/kT} - 1)^{-1} \tag{A.2}$$

represents the number of thermal longitudinal-optical phonons and

$$R_{f,o} = < \vec{P}_f^*,\vec{k}_f \; |R| \; \vec{P}_o^*,\vec{k}_o > \tag{A.3}$$

is the transition matrix element due to the scattering processes shown in Fig. 7a and 7b.

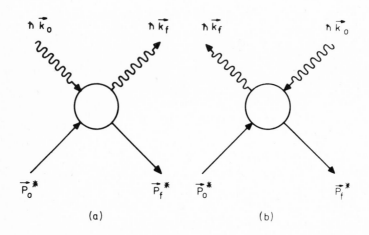

Fig. 7.   a) Resonant scattering process. b) Non-resonant scattering process.

The initial state $|0> = |\vec{p}_o, \vec{k}_o >$ consists of a polaron characterized by its momentum $\vec{p}_o$, energy $E(p_o)$, and an incoming phonon with wave vector k_o; the final state $|f> = |\vec{p}_f, \vec{k}_f >$ consists of a polaron with momentum $\vec{p}_f$ and energy $E(p_f)$ and in addition an outgoing phonon with wave vector $\vec{k}_f$.

An explicit formal expression for the R matrix is easily obtained in terms of the in-and outgoing wave operators: $a_+(k)$ and $a_-(k)$ introduced in [10] (eq.(13)) and defined as

$$a_+^\dagger(\vec{k}) = a^\dagger(\vec{k}) i - \frac{V(k)}{\hbar} \int_{-\infty}^{0} dt \, e^{-\varepsilon|t|} \, e^{i[\vec{k}\,\vec{r}(t) - \omega t]} \qquad (A.4)$$

$$a_-^\dagger(\vec{k}) = a^\dagger(\vec{k}) + i \frac{V(k)}{\hbar} \int_{0}^{\infty} dt \, e^{-\varepsilon|t|} \, e^{i[\vec{k}\,\vec{r}(t) - \omega t]} \qquad (A.5)$$

In terms of these operators, the S-matrix element for the scattering of a polaron in momentum state $\vec{p}_o$ by a phonon of momentum $\vec{k}_o$ into a final state, characterized by momenta $\vec{p}_f$ and $\vec{k}_f$ is given by

$$S_{f,o} = <\vec{p}_f^* | a_-(\vec{k}_f) a_+^\dagger(\vec{k}_o) | \vec{p}_o^* > \qquad (A.6)$$

From Eqs. (A.5) and (A.6) one immediately obtains

$$a_-^\dagger(\vec{k}) = a_+^\dagger(\vec{k}) - i \frac{V(k)}{\hbar} \int_{-\infty}^{+\infty} dt \, e^{-\varepsilon|t|} e^{+i[\vec{k}\,\vec{r} - \omega t]} \qquad (A.7)$$

so that

$$S_{f,o} = <\vec{p}_f^* | a_-(\vec{k}_f) a_-^\dagger(\vec{k}_o) | \vec{p}_o^* > - i \frac{V(k_o)}{\hbar} \int_{-\infty}^{+\infty} dt <\vec{p}_f^* \, a_-(\vec{k}_f)$$
$$\times \, e^{i[\vec{k}\,\vec{r}(t) - \omega t|} e^{-\varepsilon|t|} |\vec{p}_o^* > \qquad (A.8)$$

which leads to the familiar relation between the S- and R-matrix elements

$$S_{f,o} = \delta_{\vec{k}_f, \vec{k}_o} \, \delta_{\vec{p}_f^*, \vec{p}_o^*} - 2i\pi R_{t,o} \, \delta\{E_f - E_o\} \qquad (A.9)$$

where

$$R_{f,o} = V(\vec{k}_o) <\vec{p}_f^* | a_-(\vec{k}_f) e^{i\vec{k}_o \, \vec{r}(o)} |\vec{p}_o^* > \qquad (A.10)$$

and

$$E_j = E(\vec{p}_j^*) + \hbar\omega(\vec{k}_j) \qquad j = \{f, o\} \qquad (A.11)$$

The expression (A.1) for the inverse of the polaron lifetime becomes

$$\gamma^{-1} = -\frac{2}{\hbar} \bar{N} \, \mathrm{Jm} \left(\int d\vec{k}_o \; <\vec{P}_o^*, \vec{k}_o | R | \vec{P}_o^*, \vec{k}_o> \right) \tag{A.12}$$

We shall use relation (A.12) in order to calculate the polaron mobility. This relation leads more rapidly to the correct asymptotic behaviour for the polaron mobility in the weak coupling limit [8] than the familiar golden rule expression for γ^{-1} of Eq. (A.1). Making use of Eq. (A.7) and (A.10), the R-matrix element can also be written as follows (after time ordering of the operators involved):

$$<\vec{P}_f^*, \vec{k}_f | R | \vec{P}_o^* \vec{k}_o> = \frac{V^*(k_f) V(k_o)}{\hbar} (<\vec{P}_f^*| e^{i\vec{k}_o \vec{r}(o)} \int_{-\infty}^{0} d \; e^{-\varepsilon|\tau|} e^{-i[\vec{k}_f \vec{r}(\tau) - \omega\tau]} |\vec{P}_o^*>$$

$$+ <\vec{P}_f^*| \int_{0}^{\infty} d\tau \, e^{-\varepsilon \tau} e^{-i[\vec{k}_f \vec{r}(\tau) - \omega\tau]} e^{i\vec{k}_o \vec{r}(0)} |\vec{P}^*>) \tag{A.13}$$

Integration over time of Eq. (A.13) yields directly the Chew-Low equation for polarons:

$$<\vec{P}_f^*, \vec{k}_f | R | \vec{P}_o^*, \vec{k}_o> = V^*(k_f) V(k_o) (<\vec{P}_f^*| e^{i\vec{k}_o \vec{r}} \frac{1}{E(P_o^*) - H - \hbar\omega + i\varepsilon} e^{i\vec{k}_f \vec{r}} |\vec{P}_o^*>$$

$$+ <\vec{P}_f^* e^{-i\vec{k}_f \vec{r}} \frac{1}{E(P_f^*) - H + \hbar\omega + i\varepsilon} e^{+i\vec{k}_o \vec{r}} |\vec{P}_o^*>) \tag{A.14}$$

The first term in the right-hand side of Eq. (A.14) gives rise to a nonresonant contribution and is related to the so-called "nonresonant process" (see Fig. 7b). In the case of scattering of slow polarons,

$$E(P^*) < \hbar\omega \tag{A.15}$$

it can be seen from the optical theorem that this contribution cancels exactly, and that only the second term in the right-hand side of relation (A.14) contributes to the scattering, note (A.12). This term gives rise to a "resonant process" (see Fig. 7a), when the energy of the intermediate state E_i satisfies

$$E_i = E(P_f^*) + \hbar\omega \tag{A.16}$$

Here we concentrate on the resonant process which is the leading term for isotropic elastic scattering of slow polarons. For mobility calculations at high electric fields (large initial polaron momenta) the first term in the right-hand side of Eq. (A.14) should also be included.

The preceding remarks allow us to write

$$\frac{1}{\tau} = \frac{2}{\hbar} \bar{N} \, \text{Re} \left(\int d\vec{k} \, \frac{|V(k)|^2}{\hbar} < \bar{P}_o^*| \int_0^\infty d\tau \, e^{i(\omega + i\varepsilon)\tau} e^{-i\vec{k}\,\vec{r}(\tau)} e^{i\vec{k}\,\vec{r}(o)} |\bar{P}_o^* > | \right) \tag{A.17}$$

At this point we make use of the time dependence of the polaron position operator $\vec{r}(t)$ obtained in Eq.(27) of Ref. [9]

$$\vec{r}(t) = \vec{R} + \frac{\vec{P}^*}{m^*} t + \frac{i}{m} \sum_{\vec{k}} \vec{k}(f_{\vec{k}} \frac{\exp(-i\gamma_{\vec{k}}t)}{\gamma_{\vec{k}}} e^{i\vec{k}\cdot\vec{R}} a_+(\vec{k}) - H.c.) \tag{A.18}$$

This allows us to evaluate the expression $e^{-i\vec{k}.\vec{r}(t)} e^{i\vec{k}.\vec{r}(o)}$ if the translation and oscillation components of the electron motion are considered to be independent.

After proceeding in a similar manner as in Ref. [9] for the disentangling of both the translation and the oscillation part in Eq.(21) of [9] we are led to

$$e^{-i\vec{k}\,\vec{r}(t)} e^{i\vec{k}\,\vec{r}(o)} = \exp\left[-i(\hbar k^2/2m^*)t \right]$$
$$\times \exp[- i(\vec{k}\,\vec{P}^*/m^*)t] \, 0_1 0_2 0_3 \tag{A.19}$$

with

$$0_1 = \exp\left[\sum_{\vec{k}} (\frac{\vec{k}'\,\vec{k}}{m})^2 \frac{|\vec{f}_k|^2}{\gamma_{\vec{k}}^2} \, [\exp(-i\gamma_{\vec{k}} t) - 1] \right]$$

$$0_2 = \exp\left[i \sum_{\vec{k}'} (\frac{\vec{k}\,\vec{k}'}{m}) f_{\vec{k}}^* a_+^\dagger(\vec{k}') e^{i\vec{k}'\,\vec{R}}(\frac{\exp(i\gamma_{\vec{k}'} t) - 1}{\gamma_{k'}}) \right]$$

$$0_3 = \exp\left[-i \sum_{\vec{k}'} (\frac{\vec{k}\cdot\vec{k}'}{m}) f_{\vec{k}'} \, \frac{\exp(-i\gamma_{\vec{k}'} t) - 1}{\gamma_{\vec{k}'}} e^{i\vec{k}'\,\vec{R}} a_+(\vec{k}') \right]$$

where

$$\gamma_{\vec{k}} = \omega - (\bar{k}\,\bar{P}^*)/m^* + \hbar k^2/2m^* \tag{A.20}$$

and the function f_k is determined by the integral equation

$$f_k = \left\{ V_k \, \exp\left[- \frac{1}{2} F(k,o) \right] / \gamma_k \right\}$$
$$\times (1 - (i/\gamma_k^2)[M(o) - M^*(o) - M(-\gamma_k) + M^*(\gamma_k)])^{-1} \tag{A.21}$$

with

$$M(z) = \frac{1}{3} \sum_{\vec{k}'} \frac{k'^2 V_{k'}^2}{\hbar m} g(k', \gamma_{k'} + z)$$

$$g(k,z) = \int_0^\infty d\tau\, e^{iz\tau} e^{-\mathcal{E}|\tau|}\, \exp[\vec{F}^*(\vec{k},\tau) - \vec{F}(\vec{k},o)] \qquad (A.22)$$

$$F(k,\tau) = k^2 \sum_{\vec{k}'} \frac{k'^2}{3m^2} |f_{k'}|^2 \gamma_{k'}^{-2} \exp(i\gamma_{k'}\tau)$$

Notice that the matrix elements of (A.20) are between zero phonons, so that

$$\tau^{-1} = \frac{2}{\hbar} \, NRe \left\{ \sum_{\vec{k}} \frac{|V_k|^2}{\hbar} \int_0^\infty dt\, \exp\left[i\left(\omega - \frac{\hbar k^2}{2m^*} \pm i\mathcal{E}\right) t \right] O_1 \right\} \qquad (A.23a)$$

where

$$O_1 = \exp\left[\sum_{\vec{k}} \left(\frac{\vec{k}\,\vec{k}'}{m}\right)^2 \frac{|f_{k'}|^2}{\gamma_{k'}^2} [\exp(-i\gamma_{\vec{k}'}, t) - 1] \right] \qquad (A.23b)$$

If the electron-LO phonon coupling strength increase it is expected [1] that the electron undergoes excitations in the potential well
caused by the induced polarization. In order to simplify the numerical
computation, we approximate the time-dependent function (Eq.(A.23b)) by

$$F(k,\tau) = F(k,o)\, e^{-i\Omega_o \tau} \qquad (A.24)$$

This approximation has been justified in Ref. [9] and consists in
replacing the frequency distribution in Eq.(A.23) by one single
frequency Ω_o, the choice of which is determined variationally [9].
Making use of this approximation we find

$$\tau^{-1} = \frac{2N}{\hbar} \int d\vec{k}\, \frac{|V_k|^2}{\hbar} \, Re \left\{ \int_0^\infty d\tau\, \exp\left[i\left(\omega - \frac{\vec{k}\,\vec{P}}{m^*} - \frac{\hbar k^2}{2m^*} + i\mathcal{E}\right)\tau \right] \right.$$

$$\left. \times \exp[-F(k,o)(1-\cos\Omega_o)] \right. \qquad (A.25)$$

Eqs.(2) follows then immediately.

REFERENCES

1. J.Devreese, J. De Sitter, M.Goovaerts, Phys.Rev. $\underline{B5}$, 2367 (1972)
2. R.P.Feynman, Phys.Rev. $\underline{97}$, 660 (1955)
3. K.Thornber, Phys.Rev. $\underline{B9}$, 1929 (1971)
4. J. De Sitter, L.F.Lemmens, J.Devreese, unpublished
5. A.B.Kunz, J.T.Devreese, T.C.Collins, J.Phys. $\underline{C5}$, 3259 (1972)
   J.T.Devreese, A.B.Kunz, T.C.Collins, Solid State Commun. $\underline{11}$, 673 (1972)
6. R.P.Feynman, R.W.Hellwarth, C.K.Iddings, P.M.Platzman, Phys.Rev. $\underline{127}$, 1004 (1962)
7. D.C.Langreth, Phys.Rev. $\underline{159}$, 717 (1967)
8. J.T.Devreese, R.Evrard, E.Kartheuser, Phys.Stat.Sol.(b) $\underline{90}$, K73 (1978)
9. J.T.Devreese, R.Evrard and E.Kartheuser, Phys.Rev. $\underline{B12}$, 3353 (1975)
10. E.Kartheuser, J.Devreese, R.Evrard, Phys.Rev. $\underline{B19}$, 546 (1979)
11. L.P.Kadanoff, Phys.Rev. $\underline{130}$, 1364 (1963)
12. K.Thornber, R.P.Feynman, Phys.Rev. $\underline{B1}$, 4099 (1970)
13. F.C.Brown, in "Defects in Solids", ed. by J.H.Crawford (1970) p.491-549
14. R.C.Hughes, Phys.Rev.Lett. $\underline{35}$, 449 (1975)
15. J.T.Devreese, in "Path Integrals and their Applications in Quantum Statistical and Solid State Physics", eds. G.J.Papadopoulos and J.T.Devreese, Plenum Corp. (1978) p.315-357
16. G.Bauer, in "Springer Tracts in Moder Physics", Vol.74, Springer Verlag (New York) 1974.
17. J.Devreese and R.Evrard, phys.stat.sol.(b) $\underline{78}$, 85 (1976)
18. G.Alberga and J.T.Devreese, phys.stat.sol.(b) $\underline{90}$, K149 (1978)
19. H.Fröhlich and B.V.Paranjape, Proc.Phys.Soc. $\underline{69}$, 21 (1956)
20. H.Fröhlich, H.Pelzer, S.Zienau, Phil.Mag. $\underline{41}$, 221 (1950)
21. T.D.Lee, F.E.Low and D.Pines, Phys.Rev. $\underline{90}$, 297 (1953)
22. A.V.Tulub, Soviet Phys. JETP $\underline{14}$, 1301 (1962)
23. N.N.Bogolubov and S.V.Tjablikov, Zh.Eksp.Teor.Fiz. $\underline{19}$, 256 (1949)
24. V.M.Buimistrov and S.I.Pekar, Soviet Phys. JETP $\underline{5}$, 970 (1957)
25. V.M.Buimistrov and S.I.Pekar, Soviet Phys. JETP $\underline{6}$, 977 (1958)
26. W.Van Haeringen, Phys.Rev. $\underline{137}$, 1902 (1965)
27. D.Matz and B.C.Burkey, Phys.Rev. $\underline{B3}$, 3487 (1971)
   D.Matz, in "Polarons in Ionic Crystals and Polar Semiconductors", ed. by J.T.Devreese, North Holland (1972) p.463
28. E.Kartheuser, in "Polarons in Ionic Crystals and Polar Semiconductors", ed. by J.T.Devreese, North Holland (1972)
29. F.Low, D.Pines, Phys.Rev. $\underline{98}$, 414 (1955)
30. T.D.Schultz, Phys.Rev. $\underline{116}$, 526 (1960)
31. D.C.Langreth, Phys.Rev. $\underline{137}$, A760 (1965)
32. Y.Osaka, Progr.Tehoret.Phys. (Kyoto) $\underline{25}$, 517 (1961)
33. A.Weyland, Ph.D. Thesis (unpublished)
34. D.C.Langreth and L.P.Kadanoff, Phys.Rev. $\underline{133}$, A1070 (1964)
35. J.Irmer and P.Suptitz, Phys.Stat.Sol. $\underline{1}$, K81 (1961)
36. R.Van Heyningen, Phys.Rev. $\underline{128}$, 2112 (1962)
37. C.H.Seager and D.Emin, Phys.Rev. $\underline{B2}$, 3421 (1970)
38. F.C.Brown and N.Inchauspé, Phys.Rev. $\underline{121}$, 1303 (1961)
39. E.M.Gershenzon, V.A.Ilin, I.N.Kurilenko and L.B.Litvak-Gorskaya, Soviet Phys. - Semicond. $\underline{6}$, 1457 (1973); $\underline{6}$, 1612 (1973)

STRESS DEPENDENCE OF QUANTUM LIMIT HALL EFFECT AND
TRANSVERSE MAGNETORESISTANCE IN n-InSb

E.J. Fantner
Institut für Physik
Montanuniversität Leoben

A-8700 Leoben, Austria

ABSTRACT

The dependence of the magnetic freeze-out and the sign change of
the Hall effect on impurity concentration, compensation and temperature
has been investigated under uniaxial stress up to 4 kbar in the magnetic field range of 1-10 Tesla. Both the transverse and longitudinal
conductivity are strongly increased by an uniaxial stress of some
kbar. While the zero-stress results were described by a quantum transport theory including the effect of the nonparabolicity of the conduction band on ionized impurity scattering consistently, this theory
fails to reproduce the stress dependence of the conductivity tensor.

1. INTRODUCTION

At liquid helium temperatures the magnetic field dependence of
the Hall effect and the transverse magnetoresistance in n-InSb are
governed by the magnetic freeze-out of the conduction electrons into
localized states due to the shrinkage of the impurity wave functions
in a magnetic field [1-4]. The reduction amount of the number of free
carriers depends on the overlapping of the impurity level with the conduction band due to band tailing effects [5-7]. At magnetic fields
above about 2 Tesla depending on doping and temperature a sign change
of the Hall effect occurs, whereas the transverse magnetoresistance
saturates with increasing magnetic field. This effect was first observed by Neuringer [4]. The dependence of the magnetic field B_c where
the sign change occurs, on impurity concentration, compensation and
temperature is similar to the dependence of the magnetic freeze-out on
these parameters [6,8]. This was described by a quantum transport calculation including the effect of the nonparabolicity of the conduction
band on ionized impurity scattering [9-11]. This theory gives various
positive contributions to the conductivity tensor, dominating the
normal Hall conductivity at high magnetic fields, when most of the
free conduction electrons are frozen out. Taking into account the difference of the band parameters of InAs and InSb this theory describes

also the experimental data of n-InAs at extreme high magnetic fields [12,13].

It is the purpose of this paper to present the systematic investigation of the stress-dependence of the anomalous contributions to the Hall conductivity and to compare these results with the zero-stress experiments.

The application of uniaxial stress changes the band structure of InSb [14,15]. The stress-induced increase of the effective mass is clearly seen in an intensified magnetic freeze-out with increasing stress applied [16]. The sign change of the Hall effect is shifted to lower magnetic fields, the components of the conductivity tensor are increased by up to one order of magnitude at an uniaxial stress of some kbar [17]. The theory of Kriechbaum et al. [9-11], which describes the zero-stress results very well, fails to reproduce both the magnitude and the anisotropy of the stress-induced changes of the condunctivity tensor at magnetic fields for above the sign change of the Hall effect.

2. ZERO-STRESS EXPERIMENTS

Magnetotransport experiments on intermediate doped n-InSb at very high magnetic fields (up to 20 Tesla) were first performed by Neuringer et al. [3,4]. The observed steep increase of the Hall coefficient with increasing magnetic fields (Fig. 1) was attributed to the freeze-out of free electrons into localized donor states. The activation energies, deduced from the slope of R_H versus magnetic field, decrease with increasing impurity concentration and are always much smaller than the optically determined activation energies [7,18,19]. This can be described quantitatively by the theory of Dyakonov et al. [5], modified by using a magnetic field depended screening length [20-22].

Further increase of the magnetic field leads to a sign change of the Hall effect [4,6,19] accompanied by a saturation of the transverse magneto-resistance [6,8,19] (Figs 1 and 2). The longitudinal magneto-resistance shows a fairly similar behaviour, the absolute values are smaller by a factor of 2 to 4. As shown in Fig. 1 at magnetic fields far above the sign change the Hall coefficient approaches a constant value of about $+10^5$ cm^3/C, independent of temperature and doping [6,8].

To eliminate influences of sample inhomogeneitis or non-symmetric Hall contacts the correct Hall voltage and potential drop across the sample were always determined by averaging the measured voltages for

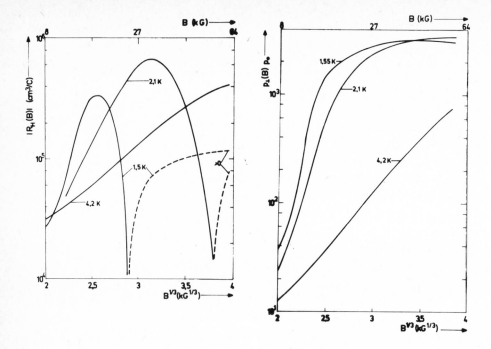

Fig. 1. Dependence of the Hall coefficient on the magnetic field with the temperature as parameter; the dashed curves indicate positive Hall coefficients (impurity concentration $N_I = 4.4 \times 10^{14}$ cm^{-3}, compensation ratio $K = 0.21$)

Fig. 2. Dependence of the transverse magnetoresistance on the magnetic field with the temperature as parameter for the sample of Fig. 1.

both polarities of current and magnetic field. As an example the magnetic field dependence of the four Hall voltages measured on a sample with very symmetric Hall contacts is shown in Fig. 3. The experimental arrangement and the sample preparation are described in detail in a previous paper [8].

For a systematic investigation of the influence of impurity concentration and compensation, samples doped by thermal neutron irradiation [23] were used in addition to samples doped in the melt. Fig. 4 shows the dependence of the magnetic field B_c, where the sign change occurs, on temperature for several samples of equal acceptor concentration. Obviously, the sign change of the Hall effect is shifted to lower magnetic fields, when the temperature is lowered and the compensation

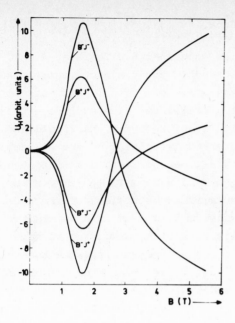

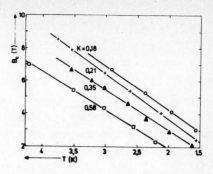

Fig. 4. Temperature dependence of the magnetic field B_c where the Hall effect changes sign. Parameter is the compensation ratio $K = N_A/N_D$ with the acceptor concentration $N_A = 0.75 \times 10^{14}$ cm^{-3} for all samples.

Fig. 3. Magnetic field dependence of the Hall voltages for both polarities of current and magnetic field at 1.8 K for the sample of Fig. 1.

is increased. Investigations on samples of equal compensation, but different impurity concentrations showed a lowering of B_c by lower impurity concentrations [8]. This means, that the systematic dependence of B_c and of the magnetic freeze-out on impurity concentration, compensation and temperature are very similar.

Several mechanisms, which might be responsible for the sign change of R_H are discussed previously [8]. The possible interpretations in terms of hopping or impurity band conduction, of free holes and p-type inversion layers have been experimentally examined. Hopping conduction can not describe the saturation of the transverse magneto-resistance with increasing magnetic field [24]. Furthermore, the observed frequency-independence of the conductivity up to 200 kHz contradicts the theory of Shklovskii [24]. An interpretation by impurity band conduction would predict a dependence on compensation opposite to the observed behaviour (compare Fig. 4). Magnetotransmission experiments were performed on pure samples with an HCN laser (337 μm) in Faraday

geometry at 4.2 K up to 8.5 Tesla, which is far above the positions of
the cyclotron resonance absorption peaks for free holes [25]. Although
the sign change of R_H in these samples occurs at 6.8 Tesla at 4.2 K,
no hole resonance absorption was observed in the range of investigation.
These magnetotransmission experiments - together with experiments with
various surface roughness and sample geometries - also exclude a possible influence of a p-type inversion layer. Recently, Mansfield and
Kusztelan [26] (refered to as MK) observed a striking effect of sample
preparation on magneto-resistance and Hall effect data in InSb. Soldering
the leads to the specimen they observed only a small increase of R_H with increasing
magnetic field with a zeroing of the Hall voltage at magnetic fields of some Tesla.
The transverse magnetoresistance saturated after an increase of about one order of
magnitude. MK attributed these results to a high conducting surface
layer dominating the transport properties of the sample, when the bulk
resistance becomes very high due to free carrier-freeze out. To eliminate
this layer they reetched the sample and spotwelded the potential probes
to the sample by condenser discharge. This means, that the sample is
locally melted at the contact area. Measuring the Hall coefficient and
the transverse resistance after this treatment, MK observed a steep
increase by several orders of magnitude with increasing magnetic field.

Although the experimental results of MK have shown no evidence
for a sign change of the Hall-coefficient [27], we have proven the
possible influence of this suggested layer on our experimental results [28]. We prepared our samples in the same way as we prepared all
our samples before. After measuring the Hall effect and the transverse
magnetoresistance, we reetched the sample while the contacts and the
thin gold wires were protected by a Pizeincover. The transverse magnetoresistance data obtained before and after this treatment are compared
in Fig. 5. Taking into account the reduction of the sample cross-
-section by reetching the sample, the transverse magnetoresistance is
not affected by this treatment as shown by the broken curve in Fig. 5.
As the sign change of the Hall effect is not affected by this sample-
-treatment either these experiments are a direct proof of the sign
change of R_H as a bulk effect. This is supported by recent experiments
of L.Gutai and F.Beleznay, who observed a minimum of the free electron
concentration at about 4 Tesla and 4.2 K by using the Van der Pauw-
method [29].

Kaufmann and Neuringer [12] have reported Hall effect data which
give evidence for a sign change of R_H in n-InAs, too. Because of the
high doping of the samples investigated, the available magnetic field
strengths ($\lesssim$ 20 Tesla) allowed only the observation of a maximum of R_H.

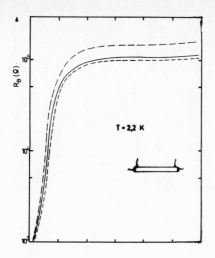

Fig. 5.
Observed transverse magnetoresistance R_B vs. magnetic field;
solid curve: re-etched without
changing contacts; broken curve:
corrected for reduced cross
section.

3. STRESS-EXPERIMENTS

The application of uniaxial compressional stress induces two
pronounced changes of the energy band structure of InSb [15]. When
uniaxial compressional stress is applied along a [100] or [111] crystallographic axis, the four-fold degenerate Γ_8-bands are split into
two doubly degenerate states with $M_J = \pm 1/2$. The bottom of the Γ_6-
conduction band is shifted up for a compressive stress by an amount of
$C_1\mathcal{E}$, where C_1 is the conduction band deformation potential and $\mathcal{E}$ is
the trace of the strain tensor. As the shift of the center of gravity
of the valence band with $J = 3/2$ can be described by the hydrostatic-
-pressure deformation potential a, the change of the energy gap with
compressive stress is given by $(a+C_1)\mathcal{E}$. The additional changes in the
effective masses of the bands should be shown in an increase of the
activation energy of the donor states. This means an intensified magnetic freeze out of electrons with increasing applied uniaxial stress
(Fig. 6). Therefore, the classical Hall conductivity term proportional
to the free carrier density should be dominated by the positive contributions to the Hall effect at lower magnetic fields than without
stress. The stress-induced shift of B_c to lower values is shown in
Fig. 7a. Additionally, an increase of R_H far above the sign change is
observed. In the magnetic freeze-out regime the transverse magnetoresistance shows a steeper increase as the applied stress increases,
whereas the saturation value decreases with stress [17] (Fig. 7b).
Accordingly to investigate the stress dependence of the anomalous Hall
effect alone, the experiments have to be performed at magnetic fields
far above the sign change.

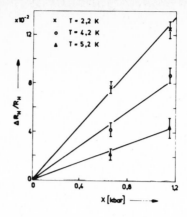

Fig. 6. Stress induced change of the Hall coefficient within the magnetic freeze out regime for various temperatures. (N_I = 2.2x10^{14} cm^{-3}, K = 0.58)

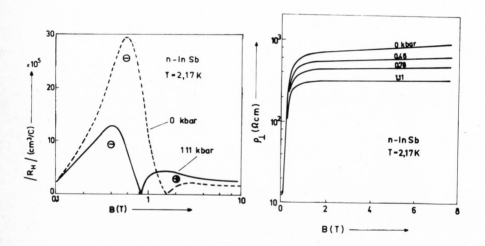

Fig. 7. Absolute value of the Hall coefficient R_H (7a) and transverse magnetoresistance $\rho_\perp$(7b) vs. magnetic field at T = 2.17 K for the sample of Fig. 6. Parameter is the applied uniaxial stress.

Both the Hall effect and the transverse magnetoresistance have been investigated at magnetic fields of 8 and 9 Tesla at 2.17 K up to an uniaxial stress of 4 kbar. The uniaxial stress has been applied parallel to the long axis of the sample clamped between a movable and a fixed piston [30]. The directions of the uniaxial stress applied have been the [001], [110] and [111] crystallographic axes, the corresponding perpendicular directions of the magnetic field have been [110], [001] and [112].

Neglecting the stress-induced anisotropy of the effective mass, the components of the conductivity tensor can be calculated from the Hall effect and transverse magnetoresistance data by eqs. (1) and (2).

$$\sigma_{xy}(B_1) = \frac{\rho_\perp(B_1)}{\rho_\perp^2(B_1)+[R_H(B_1)\cdot B_1]^2} \tag{1}$$

$$\sigma_{xy}(B_1) = \frac{\sigma_{xy}(B_1)}{\rho_\perp(B_1)} - \sigma_{xy}^2(B_1) \tag{2}$$

$\rho_\perp(B_1)$ is the specific transverse magneto-resistance at a magnetic field B_1. As an example Fig. 8 shows the stress-dependence of σ_{xy} and σ_{xx} of a very pure sample ($N_I = 4 \times 10^{14}$ cm^{-3}). The stress was applied

n-InSb   T = 2,17 K   B = 8 T   j ‖ X ‖ [111]   B ‖ [1$\bar{1}$0] ... o
[11$\bar{2}$] ... x

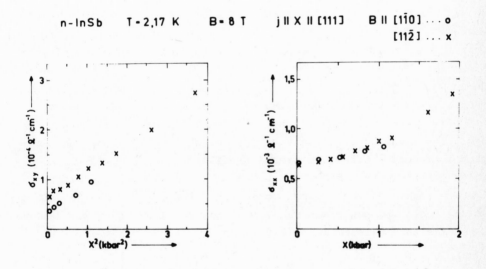

Fig. 8. σ_{xy} vs. X^2 (8a) and σ_{xx} vs. X (8b) at B=8 Tesla and T=2.17 K ($N_I = 2.2 \times 10^{14}$ cm^{-3}, $K = 0.58$); crosses: $\vec{J}\,\|\vec{X}\,\|$ [111], $\vec{B}\,\|$ [11$\bar{2}$]; circles: $\vec{J}\,\|\vec{X}\,\|$ [111], $\vec{B}\,\|$ [1$\bar{1}$0]

along a [111] crystallographic axis, the magnetic field in a [1$\bar{1}$0] (circles) and [11$\bar{2}$] (crosses) direction, perpendicular to the applied stress.

The variation of σ_{xy} with applied stress can be described by a
quadratic dependence, whereas σ_{xx} is better described by a linear dependence on stress. The stress induced anisotropy is very pronounced.
The stress-dependeces of σ_{xx} and σ_{xy} are stronger by more than a
factor of three for the orientation $\vec{X} \parallel \vec{j} \parallel$ [001] than for
$\vec{X} \parallel \vec{j} \parallel$ [111] or [110]. A dependence of σ_{xx} and σ_{xy} on different
orientations of the magnetic field could not be resolved. The differ-
ence of the absolute values of σ_{xy} for $\vec{B} \parallel$ [1$\bar{1}$0] and [11$\bar{2}$] is within
the experimental error due sample inhomogeneities. The slopes of the
σ_{xx}- and σ_{xy}-curves versus stress of other samples with the same
orientation but from different crystals always agreed with the curves
shown in Fig. 8 within 10 %. The absolute values varied by less than a
factor of two. The applied stress was always determined within an
error of 5 %.

4. THEORY

The anomalous Hall effect [31] in semiconductors arises from the
influence of the non-parabolicity of the conduction band on multiple
scattering of electronsat charged impurities. Including the vector potential of an arbitrary homogenious magnetic field in the canonical momentum [32,11] , a $\vec{k}.\vec{p}$-perturbation theory [32,33] was used to calculate
the anomalous contributions to the Hall current. Two of these anomalous
contributions were observed by an ESR-experiment at low magnetic
fields in n-InSb and n-Ge by Chazalviel and Solomon [34,35] . A
quantizing magnetic field gives rise to the freeze out of the electrons into donor states, Boltzmann statistics become applicable.
Therefore, the screened Coulomb potential is drastically affected. For
the calculation of the Coulomb potential Wallace [20] used the Lindhart
formula replacing the plane waves by Landau wave functions. For classical
statistics this calculation yields a potential strength described by [9]

$$V = - \frac{k_B T}{n} \tag{3}$$

k_B is the Boltzmann constant, T the temperature and n the density of
free electrons. As n decreases rapidly with increasing magnetid field
due to magnetic freeze out, the potential strength increases by the
same amount. As a calculation of the impurity scattering within the
Born approximation takes into account only terms quadratic in the impurity potential V, a t-matrix approximation was used to calculate the
scattering of conduction electrons at an impurity potential given by

$$v(\vec{r}) = \sum_{\vec{r}_i} v.\delta(\vec{r}-\vec{r}_i) \tag{4}$$

$\vec{r}_i$ describes the position of the ionized impurities. Using the Kubo formula all essential anomalous contributions due to the nonparabolicity of the conduction band have been proven to be linear in the density of scattering centers N_S and the strength of the scattering potential V. Consequently the anomalous Hall current j_{an} is independent of the density of the conduction electrons as shown by

$$j_{an} \propto n.N_S.V = -N_S.k_B.T \tag{5}$$

The magnetic field range, where the assumptions described above are valid, is shown for n-InSb and n-InAs by the solid lines in Fig. 9. As the magnetic fields used in the experiments are well within this range of validity, the sign change of the Hall effect can easily be interpreted within the framework of this theory. The Hall effect will change its sign at a critical magnetic field B_c, where the anomalous Hall conductivity - it is nearly independent of the magnetic field in compensated semiconductors at high magnetic fields, because N_S is always fairly equal to

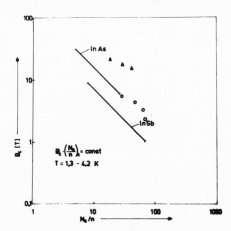

Fig. 9. Calculated dependence of the magnetic field B_c, where the sign change of the Hall effect occurs on the ratio of the density of ionized scattering centers N_S and conduction electrons n (solid lines). The experimental data for n-InAs (triangles) are taken at the magnetic field B_M, where the Hall coefficient shows its maximum.

the number of acceptors N_A - dominates the normal Hall conductivity, which decreases rapidly with increasing magnetic field due to carrier freeze out.

The main anomalous contributions to the conductivity tensor are given by [11]

$$\sigma_{xx}^{an} = - \frac{e^2 N_s \hbar}{2m^+ E_1} \left(\frac{k_B T}{\hbar \omega_c^+}\right)^{3/2} \cdot \left(1 + \frac{\nu E_1}{E_2}\right) \tag{6}$$

$$\sigma_{xy}^{an} = \frac{e^2 N_s \bar{\hbar}}{m^+ E_1} \left(1 + \frac{\nu E_1}{E_2}\right) \tag{7}$$

m^+ is the effective mass of the electrons, $\bar{\hbar}$ the Planck constant devided by 2π, ω_c^+ is the cyclotron frequency. ν describes the spin splitting of the Landau levels (for InSb $\nu \simeq -\frac{1}{3}$) and E_1 and E_2 are energy parameters. For InSb the value of the spin orbit splitting Δ is much larger than the forbidden energy gap $\mathcal{E}_g$. Therefore, E_1 and E_2 equal $\mathcal{E}_g$ in value. For InAs, E_1 and E_2 are about 0.5 eV and 0.3 eV.

As discussed in chapter 3, uniaxial compressional stress changes the wave-functions and the band structure of InSb and as a result, the anomalous Hall effect will be affected too. The stress dependence of σ_{xy} was calculated by Biernat and Kriechbaum [36] using the stress Hamiltonian given by Pollak and Cardona [37].

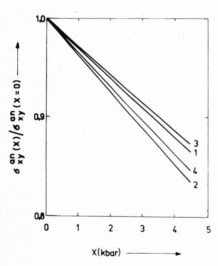

Fig. 10. σ_{xy}^{an} as a function of $\vec{X} \parallel [001]$ (curves 1 and 2) and $\vec{X} \parallel [111]$ (curves 3 and 4). Curves 1 and 3 have been calculated for a hydrostatic pressure deformation potential a = 5.5 eV, curves 2 and 4 for a = 6.5 eV.

For the stress applied in [001] - and [111] -crystallographic directions this Hamiltonian was exactly diagonalized at the Γ-point ($\vec{k} = 0$). With these new eigenvalues and eigenfunctions the $\vec{k}.\vec{p}$-perturbation theory was performed analogously to the calculation for stress equal zero. InSb terms proportional $\underset{\sim}{\xi}.\vec{k}$ and $\underset{\sim}{\xi} \vec{k}^2$ are very small [14] so that they were neglected. As shown in Fig. 10 this calculation predicts a decrease of the anomalous Hall effect for all directions of applied uniaxial stress.

5. DISCUSSION

It was shown in previous papers [6,8,9], that this model calculation is able to explain the dependence of the critical magnetic field B_c on doping, compensation and temperature. As an example, Fig.9 shows the dependence of B_c on the ratio of the density of scattering centers and the density of free electrons for n-InSb and n-InAs. The slope of the experimentally observed decrease of B_c with decreasing electron concentration - this means experimentally a lower temperature - is described very well by this theory. The fact that the experimentally determined Hall angle is always considerably smaller than predicted by these calculations might be due to the use of δ-function to describe the impurity scattering potential. Recently, Ishida and Otsuka [38] have suggested, that at high magnetic fields the Hall effect in pure n-InSb will be dominated by electrons in the conduction band, the transverse magneto resistance by activated impurity conduction. This would give a further argument for a Hall angle smaller than predicted by the theory.

As the stress induced changes of the band parameters are within some percent for an uniaxial stress experimentally available, the theory described above cannot reproduce the observed increase of the components of the conductivity tensor. Although the influence of higher conduction bands should affect the calculated stress dependence [17] of the anomalous Hall effect, the predicted effect remains still far too small compared to the experimental results.

Porowsky et al. [39] have found experimental evidence for donor ions occupying non-equivalent lattice positions in InSb. If neither magnetic field nor stress are applied the energy levels of these states should be about 100 meV above the conduction band minimum at the Γ-point, their pressure coefficients are close to the values for the X- and the L-valley. Considering the shift of the Γ-minimum due to the magnetic field and the uniaxial stress, an influence of these metastable states on the observed transport phenomena cannot be excluded without an experimental proof.

ACKNOWLEDGEMENTS - It is a pleasure to acknowledge F.Kuchar, with whom most of the work presented here was performed. I am indepted to R.Danzer for many clarifying discussions on the theory of the anomalous Hall effect. I thank Prof. G.Bauer for his continuous interest and support of this work, H.Biernat, M.Kriechbaum and R.Danzer for communating their results prior to publication. This work was supported by the "Fonds zur Förderung der Wissenschaftlichen Forschung in Österreich" (project No.2785).

REFERENCES

1. Y.Yafet, R.M.Keyes and E.M.Adams, J.Phys.Chem.Solids 1, 137 (1956)
2. R.W.Keyes and R.J.Sladek, J.Phys.Chem.Solids 1, 143 (1956)
3. O.Beckman, E.Hanamura and L.J.Neuringer, Phys.Rev.Letters 18, 773 (1967)
4. L.J.Neuringer, Proc. 9th Int.Conf.Phys.Semicond. (ed. S.M.Ryvkin, Nauka, Moscow (1968), Vol.2, p.715)
5. M.I.Dyakonov, A.L.Efros and D.L.Mitchell, Phys.Rev. 180, 813 (1969)
6. E.J.Fantner, F.Kuchar, G.Bauer, M.Kriechbaum and H.Biernat, Proc. 12th Int.Conf.Phys.Semicond. (ed. M.Pilkhun, Teubner, Stuttgart (1974) p.249)
7. F.Kuchar, E.J.Fantner and G.Bauer, J.Phys.C: Solid State Phys. 10, 3577 (1977)
8. E.J.Fantner, F.Kuchar and G.Fauer, Phys.Stat.Sol.(b) 78, 643 (1976)
9. H.Biernat and M.Kriechbaum, Phys.Stat.Sol.(b) 78, 653 (1976)
10. H.Biernat, Thesis, Graz (1977)
11. R.Danzer and M.Kriechbaum, Acta Phys. Austr. (in press)
12. A.Kaufmann and L.J.Neuringer, Phys.Rev. B2, 1840 (1970)
13. R.Danzer, E.J.Fantner and F.Kuchar, submitted to Phys.Stat.Sol.(b)
14. G.Bir and G.Pikus, Sov.Phys.-Solid State 3, 221 (1962)
15. D.G.Seiler, Int. Conf. "The Application of High Magnetic Fields in Semiconductor Physics" (ed. G.Landwehr, Würzburg 1974) p.492
16. E.J.Fantner and F.Kuchar, submitted to Phys.Stat.Sol.(b)
17. F.Kuchar, E.J.Fantner and M.Kriechbaum, Proc. 14th Int.Conf.Phys. of Semicond. (ed. B.L.H.Wilson, Institute of Physics Conference Series Number 43, Edinburgh 1978), p.549
18. E.Gornik, Phys.Rev.Letters 28, 595 (1972)
19. E.J.Fantner and G.Bauer, Solid State Commun. 13, 629 (1974)
20. P.R.Wallace, J.Phys.C: Solid State Phys. 7, 1136 (1974)
21. S.D.Jog and P.R.Wallace, J.Phys.C: Solid State Phys. 8,3608 (1975)
22. E.J.Fantner and F.Kuchar, submitted to J.Phys.C: Solid St. Phys.
23. F.Kuchar, E.J.Fantner and G.Bauer, Phys.Stat.Sol.(a) 24,513(1974)
24. B.I.Shklovskii, Soviet Phys. Semicond. 6, 1053 (1973)
25. K.J.Button, B.Lax and C.C.Bradley, Phys.Rev.Lett. 21, 350 (1968)
26. R.Mansfield and L.Kusztelan, J.Phys.C: Solid State Phys. 11, 1 (1978)
27. R.Mansfield, private communication
28. E.J.Fantner and F.Kuchar, J.Phys.C: Solid St. Phys.(in press)
29. L.Gutai and F.Beleznay, private communication
30. F.Kuchar, K.Veigl and E.J.Fantner, Rev.Sci.Inst. 50(2),245 (1979)
31. J.M.Luttinger, Phys.Rev. 112, 739 (1958)
32. P.Nozieres and C.Lewinier, J.Physique 34, 901 (1973)
33. R.Danzer and M.Kriechbaum, Acta Phys. Austr. 47, 265 (1977)
34. J.N.Chazalviel and I.Solomon, Phys.Rev.Lett. 29, 1676 (1972)
35. J.N.Chazalviel, Phys. Rev. B11, 3918 (1975)
36. H.Biernat and M.Kriechbaum, Acta Phys. Austr. (in press)
37. F.H.Pollak and M.Cardona, Phys.Rev. 172, 816 (1968)
38. S.Ishida and E.Otsuka, J.Phys.Soc.Japan 46, 1207 (1978)
39. S.Porowski, M.Konczykowski and J.Chroboczek, Phys.Stat.Sol.(a) 63, 291 (1974)

PHOTOLUMINESCENCE IN AMORPHOUS SEMICONDUCTORS

I. Kósa Somogyi, M. Koós and V.A. Vassilyev*
Central Research Institute for Physics
H-1525 Budapest, P.O.B. 49, Hungary

1. INTRODUCTION

Excited species can discard their excess (above thermal) energy
either radiatively by emission of photons or non-radiatively by phonon
emission. During photoluminescence (PL), excess energy is supplied by
means of incident light quanta absorbed by the system, followed by desexcitation occuring as emission of photons. Emitted light quanta are
usually of smaller energy than the incident ones, indicating that non-
-radiative processes intervene between the acts of excitation and desexcitation, transferring part of the absorbed energy to increased
lattice vibrations. During the non-radiative interval between photon
absorption and emission, thermalization and network relaxation can take
place. PL, an interplay of radiative and non-radiative processes, by
its numerous measurable parameters, e.g. characteristics of the excitation and PL spectra, the change in PL intensity with temperature,
composition, annealing, its decay kinetics, fatigue during continuous
excitation, the effect of doping on these processes, etc., PL can give
valuable information on the band structure of the material and on the
density of localized states within the forbidden gap. The last factor
is especially valuable in the case of amorphous semiconductors (a-SCs),
where band tails and states in the gap seem to play a decisive role in
determining optical and electrical properties.

PL as a useful method for the study of amorphous semiconductors
has been gaining pace recently. Since the first observation of PL in
chalcogenide glasses by Kolomiets et al. [1] several groups have carried
out extensive measurements on different chalcogens and chalcogenide
systems and have tried to rationalize their findings by comparing them
with results obtained by other methods. In this respect the combined PL
and ESR measurements of Bishop et al. [2] are especially valuable.

Extensive reviews by Street [3,50] describe the PL properties of
amorphous semiconductors and endeavour to explain them in terms of the

* A.F.Ioffe Physico-Technical Institute, Leningrad, USSR

Mott-Street model supposing a relatively high density (10^{17}-10^{18}cm^{-3}) of charged localized states in the gap. A recent excellent review of Fisher [4] summarizes the essentials of radiative recombination processes, mainly those occuring in a-Si, and interpreting them as transitions between electron-hole pairs held together by Coulomb attraction.

The present paper is a review of the PL properties of chalcogenide a-SCs as deduced mainly from data obtained on the Ge-Se system. Discussions extended somewhat beyond that system are based on measurements of other chalcogenides. First excitation, then PL is described, including the discussion of PL intensity, decay kincetics, fatigue and quantum efficiency. Existing models and data are compared, discrepancies are pointed out and a model based on self-trapping of the excited carriers is suggested.

2. <u>EXCITATION</u>

2.1 <u>The absorption process</u>

Absorption spectra represent the convolution of the densities of states in the two bands involved in the optical transition, modified by transition probabilities. Optical electronic spectra and theoretical treatments show that a pseudo-gap with low density of states separates the filled valence band from the empty conduction band in amorphous semiconductors. Mott assumed [5] that potential fluctuations arising from disorder lead to a blurring of the band edges and to the formation of localized tail states extending deep into the forbidden gap. Disorder e.g. wrong bonds, changes in connectedness of the constituent atoms, bond length variations and distorted bond angles, the presence of defects, such as voids, dangling bonds, impurities etc. would diminish the gap and increase the band tailing in such a way that most of the tailing would arise from the defects.

The forbidden gap, being ultimately a measure of the bond strength by its variation affords a means to check whether a particular method of preparation, annealing, irradiation, etc. will strengthen or weaken the bonding between the constituent atoms. It is a common finding that decreasing the deposition rate or increasing the substrate temperature in the case of evaporation and sputtering, annealing near T_g, or subsequent irradiation of the as prepared materials improve the quality of the amorphous samples whose properties approach those of the "ideal" amorphous solid in which at least conceptually no extrinsic disorder exists [6].

The absorption spectra of amorphous solids, even those with nearly ideal structure, are different from the spectra of the corresponding crystalline ones in respect both of their shape and of their energy position. Although it might be tempting to deduce absorption spectra of amorphous materials by smearing out finely structured spectra of the corresponding crystals into a smooth envelope, this procedure could lead to serious errors since absorption spectra of amorphous materials are usually significantly shifted to lower energies and transitions in the amorphous and crystalline phases can often be different in nature depending on the different energy levels.

The absorption spectra of chalcogenide glasses for weakly absorbed light in the energy range characterized by $\alpha < 10^4$ cm^{-1} always has an exponential absorption edge, the slope of which does not change much for different materials. This edge, found also in crystalline semiconductors, is called the Urbach tail and is most likely due to a broadening of excitonic transitions by random internal electric fields. The extent of this tail can be substantially reduced by treatments leading to the elimination of extrinsic disorder (e.g. voids, dangling bonds, impurities etc.).The formidable effect of H atoms on the density of tail states in amorphous Si can be explained by compound formation, since otherwise as an addition, it would greatly increase the extrinsic disorder.

The photo excitation process in chalcogenides involves the topmost filled and the lowest empty bands depicted for pure Se, Ge, GeSe$_2$ and GeSe$_4$ in Fig. 1. The lowest energy electronic transitions are between the non-bonding p-type lone pair band and the empty antibonding band. The selection rule permits only transitions involving change of the dipole moment (s-p transitions).To comply with this rule in pure Se, as a means of explaining the experimental optical spectrum, one should invoke a 20 % coupling of the non-bonding and p-type bonding band. In GeSe$_2$ the lowest energy

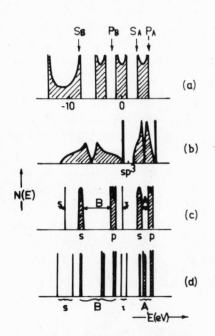

Fig. 1. Density of states of Ge$_x$Se$_{1-x}$.
(a) pure Se: the arrows indicate the localized levels introduced by Ge impurities; (b) pure Ge; (c) GeSe$_2$; (d) GeSe$_4$ [7].

electronic absorption peak is due to a transition between the p-type non-bonding lone pairs and the s-type empty antibonding band. In GeSe$_2$ the non-bonding pairs are strongly coupled with the p-type bonding states [7].

2.2 Optical gap

It is a well known distinguishing feature of amorphous chalcogenides that PL can be excited only with photons having energy equal to or greater than the band gap E_g(opt), the optically measurable forbidden band. Radiation with light quanta of smaller energy can not, as a rule, cause luminescence.

In the region of strong absorption characterized by $\alpha > 10^4$ cm^{-1} the absorption of the majority of a-SCs follows the law

$$\alpha (\hbar\omega) \propto (\hbar\omega - E_g(opt))^{1/2} \tag{1}$$

The corresponding direct allowed transitions take place between parabolic shaped bands near their extrema and separated by an energy gap of E_g(opt), the value of which can be deduced from absorption data arranged to suit eq.(1). E_g(opt) deduced in this way is called the optical gap and represents the energy separation between the extended states in the two bands. The comparision of this gap with that of the corresponding crystalline material is but of limited value since the transitions involved may be of a quite different nature in the two phases.

Instead of defining E_g(opt) by means of eq.(1) many researchers deduce it somewhat arbitrarily from the low energy part of the absoprtion spectrum [8]. The determination of E_g(opt) in this way encounters difficulties since the absorption edge of the amorphous semiconductor materials has a tail whose magnitude depends on the purity of the material and the extent of disorder. The absorption in the Urbach tail increases exponentially with excitation energy: $\alpha(\omega) = B \exp(\omega/E_e)$.

E_g(opt) is defined sometimes in a similarly arbitrary way as the excitation energy corresponding to $\alpha = 1$, $\alpha = 10$, $\alpha = 100$ cm^{-1} - depending on the choice of the researcher. The log α vs E curve is usually very steep, thus the difference in E_g(opt) caused by this indefiniteness for many cases is not too significant.

E_g(opt) for two-component systems like Ge-Se seems to be an additive property; a weighted sum of E_g(opt) for the components (at least in those composition regions where no stoichiometric compound formation occurs (see e.g. Fig. 2)). The somewhat contradictory data pertaining to the GeSe system agree in one respect: E_g(opt) for GeSe$_2$ is a local maximum. Further, the tendency of E_g(opt) to diminish by

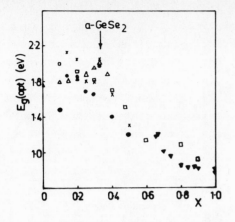

Fig. 2a. Variation of the optical energy
gap with x in Ge$_x$Se$_{1-x}$ at room
temperature. ● and x [11], △[8],
□[9], ▼ [12].

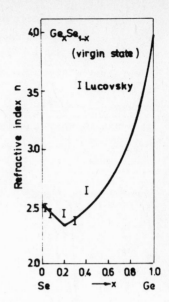

Fig. 2b. Refractive index n as a function of atomic concentration
x for the Ge$_x$Se$_{1-x}$ system. The
solid curve indicates the calculated value, bars represent
the experimental data obtained
by Lucovsky [54].

the increasing Ge content seems to
hold, although a minimum at Ge$_3$Se$_2$
was found in ref. [9]. Compound
GeSe formation is thought to be
very probable [10] even though no
corresponding local maximum on the

E$_g$(opt) vs composition curve has been detected yet. Calculations of
the refractive index as a function of x for the Ge$_x$Se$_{1-x}$ system [54]
also show the minimum to correspond roughly to GeSe$_2$ (Fig. 2b).

Annealing of the amorphous samples results in the decrease of
the Urbach tail and the widening of the E$_g$(opt) [6]. The effect - connected with the ordering of the lattice - is less pronounced in the
composition region near to GeSe$_2$ [11], but can cause an 0.5 eV widening
of the gap at other compositions, both richer and poorer in Ge content.
The strong dependence of E$_g$(opt) on structure - which can be strongly
different even for the same composition - can partly explain the
scattering of PL results obtained in different laboratories.

Since the first step leading to PL in amorphous chalcogenides is
a band to band excitation, any process causing ordering in the lattice
(annealing, compound formation, charge compensation in the tail states,
etc.) should lead to a narrowing of the PL band, provided the energy
distribution width of the recombination centres is less than those of
the tail states. Moreover, sharp edges are favourable for increasing PL
efficiency and intensity.

2.3 Excitation spectrum

Excitation spectra show the dependence of PL intensity on the energy of exciting light quanta. At first glance the excitation spectrum seems to have a Gaussian shape, but a $logI_{PL}$ vs E_x plot reveals that its low energy side is a result of at least two different absorption processes (Figs 3 and 4) [13,14,15]: the first part of the curve corresponds to transitions involving the states in the Urbach tail, the second part at higher energies is due to the basic band to

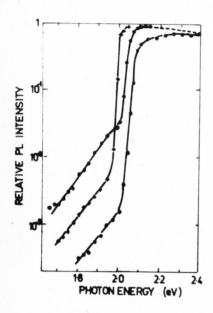

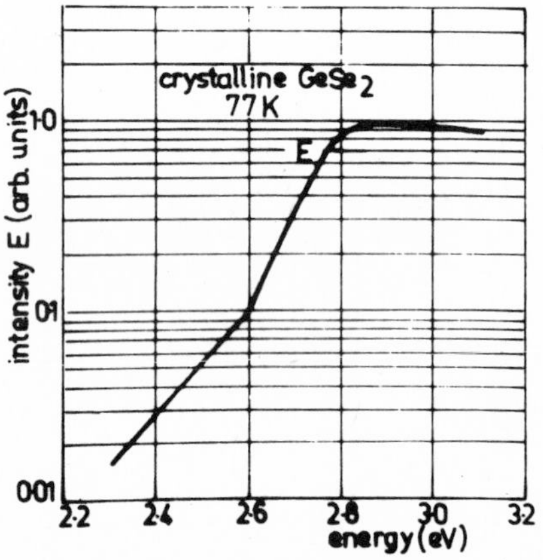

Fig. 4. Excitation spectrum for single-
-crystalline GeSe$_2$ [14]

Fig. 3. Excitation spectra of crystalline
As$_2$Se$_3$ (+), sample thickness
0.17 mm; also crystalline
(85 As$_2$Se$_3$: 15 As$_2$S$_3$) (●); upper
curve, sample thickness 0.25 mm,
lower curve 0.008 mm [3].

band transitions. It is common finding in many chalcogenides that the maximum of the spectra E_x^{max}, i.e. the excitation energy causing the most intense PL, corresponds to light quanta for which the absorption coefficient α is around 100 cm^{-1}. In everyday use this value has become a somewhat magic number for a-SCs, even though the measured values for α in fact span the range of 10-1000 cm^{-1}. These α values clearly show that excitation of PL is a bulk process.

The respective shapes of the excitation spectrum on the low and high energy sides are controlled by different processes. The principal feature of the low energy side is governed by the continuous increase in α reflecting the corresponding increase in the density of tail states and the transition probabilities.

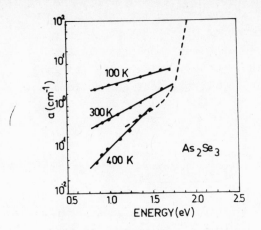

Since the density of tail
states increases during excitation [16] (Fig. 5), it is expected that the low energy side
of the excitation spectrum should
broaden and move towards lower
energies with excitation time.
No such results have been reported yet.

These induced states in the
gap are most probably due to
captured charge carriers and
their absorption spectrum in the
gap can be taken as a measure of
the distribution of trap depths.
The accumulation of these states
during excitation in the upper
half of the gap should lead to

Fig. 5. The spectral dependence of photo-
stimulated absorption coefficient
in vitreous As_2Se_3. (The dashed
curve of α is taken from ref. [17]
for T = 293 K) [16].

the possibility of exciting PL by light quanta of progressively lower
energies - ultimately much less than E_g(opt).

The high energy side of the excitation spectrum indicates that
the effectivness of higher energy photons in exciting PL decreases
steadily (Fig. 6). Since the overall absorption increases in that range

the high energy
side of the curve
shows that non-
-radiative de-
activation gets
more pronounced.
At high α, when
all the photons
are absorbed in a
very thin surface
layer, a high con-
centration of ex-
cited carriers and

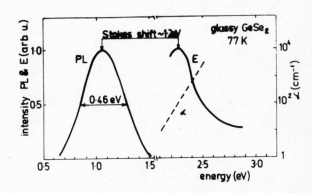

Fig. 6. PL, excitation and absorption spectra
for glassy $GeSe_2$ [14].

a high surface recombination rate are expected. Street argues [3,15]
that this process cannot be significant at α below 10^5 cm^{-1} since the
mobility in chalcogenides is not sufficient for the charge carriers
to travel the 2-3 μm distance necessary to reach the surface by diffusion. He proposes instead that only those photons will lead to PL that
are absorbed sufficiently close to the appropriate recombination

centres and that the energy dependence of absorption $\alpha_c(E_x)$, for
these centres, is necessarily weaker than that for the total absorption
$\alpha(E_x)$. The origin of the weaker dependence of α_c on energy is the
high electric field around the charged dangling bonds, the latter
having a central role in their suggested model. This Coulomb field
will broaden the absorption edge of such microregions and so lessen
its energy dependence.

3. THE EMISSION PROCESS

3.1 E_{PL}^{max}, the energetic position of the PL peak

A characteristic feature of the PL spectra of amorphous semiconductors is a symmetric Gaussian-shaped broad band centred at or near
Eg/2, the half band gap. The bands are usually devoid of fine structure with an instrumental resolution of 0.01 eV although a shoulder on
the spectra of Ge-Se containing glasses was observed when their composition differed from GeSe$_2$ [18,19]. E_{PL}^{max} for some chalcogenides
seems to be connected with the band gap energy (Fig. 7), though there
are also exceptions [3].

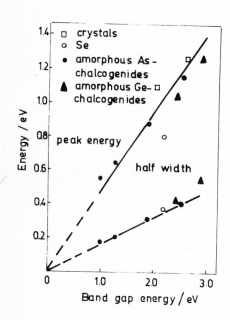

E_{PL}^{max} for the crystalline
counterparts of the chalcogenide
glasses are situated also at about
Eg/2 and the shape of their PL
spectra is very similar to that of
the glasses. The band gap for
crystals is usually larger than E_g
for their amorphous forms (both
glassy and evaporated), thus E_{PL}^{max}
of the crystals are shifted towards
higher energies: e.g. c- and
a-GeSe$_2$ E_{PL}^{max} values are at 1.17
and 1.07 eV respectively (compare
Figs 6 and 8) [14]. The 0.1 eV
difference is probably due to the
smaller band tail of the crystal.

It is to be noted that E_{PL}^{max}
in the crystalline and amorphous
forms of GeSe$_2$ are excited by
photons of 3 and 2.27 eV respectively, the corresponding α being
10^4 and 10^2 cm^{-1}, i.e. two orders
of magnitude greater in the crystal

Fig. 7. Dependence of the peak energy
and half width of the PL
spectrum on E_g(opt) in the
As- and Ge-chalcogen systems
[3,4,14].

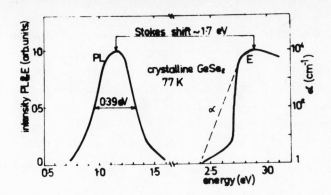

Fig. 8. PL, excitation and absorption spectra for single-crystalline GeSe$_2$ [14].

than in the glass. If the similarity of PL spectra for c- and a-forms is considered to indicate that basic processes leading to PL are approximately the same in both phases, then the above mentioned difference in α values has to be related to considerable differences in the densities of states for the amorphous and crystalline phases, unless transition probabilities are also strongly different.

E_{PL}^{max} shifts toward lower energies as the temperature increases. The shift of E_{PL}^{max} in As$_2$S$_3$ is 0.19 eV when the temperature is increased from 10 to 149 K [13] which gives a rate of 1.37×10^{-3} eV/K (roughly an order of magnitude greater than the temperature shift of E_g for such types of compounds being $\sim 4 \times 10^{-4}$ eV/K).

With regard to the fast PL decay experiments on As$_2$Se$_3$ [20] and As$_2$S$_3$ [52] when the process is studied on a nanosec time scale, E_{PL}^{max} increases with excitation energy E_x (Fig. 9). This shift of E_{PL}^{max} has not been detected in experiments with longer excitation times. This shift might be explained if one supposes that the lattice structure and concomitantly the traps would relax in time, and that this process

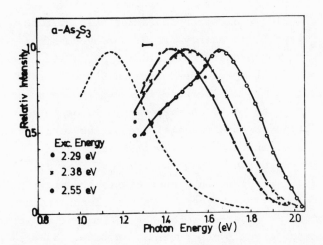

Fig. 9.

PL spectra of fast decay in a-As$_2$S$_3$ at 4.2 K [20]. Dashed line shows the result by Kolomiets et al. [21].

needs some time to be completed. During nanosec pulses the relaxation of the matrix in the vicinity of trapped charges starts but is far from being complete thus the shift of E_{PL}^{max} can be ascribed to radiative recombination occurring between non-relaxed electron-hole pairs. As an alternative explanation Shah and Bösch propose [52] that states above the mobility edge are at least partially localized and this localization impedes rapid thermalization of the excited carriers.

The addition of Tl to $GeSe_2$ and $GeSe_3$ up to 10 at% had no effect on the shape of the PL spectra and on E_{PL}^{max}, indicating that Tl atoms are not involved in the transitions bringing about photon emission [19].

Doping the chalcogenides with suitable additives can shift the PL band and/or cause the appearance of new peaks. The doping of As_2Se_3 with 0.01 at% In shifts the peak of the PL band by 0.13 eV toward smaller energies, whereas the doping with Ge results in the decrease of PL intensity and the appearance of a second peak [22]. The appearance of a second PL peak was also observed in glassy As_2S_3 doped diffusionally with Ag [53]. The second peak was assigned by the authors to the complex $(Ag^+ + dangling bond)$.

3.2 σ_{PL}, the half width

The half width of the PL spectra σ_{PL} is generally about 0.3-0.4 eV and increases with increasing temperature. According to Street's proposal [3] it is given by

$$\sigma_{PL} = 2(2W\hbar\omega\,\ell n2)^{1/2} \qquad (3)$$

where 2W is the Stokes shift, $\hbar\omega$ is the energy of lattice vibration. In view of the large Stokes shift - a common characteristic of amorphous semiconductors - σ_{PL} is defined mainly by 2W. Moreover, since $\hbar\omega$ does not change much with temperature only limited narrowing of the PL band is expected when the temperature decreases.

Stokes shift in c-$GeSe_2$ is 0.5 eV greater than that in the amorphous form [14], thus according to eq.(3) a larger σ_{PL} is expected which is not confirmed by the experimental data, even if we take into consideration that the disorder in the amorphous phase can cause broadening of the PL band by affecting the energy distribution of states in the gap. Moreover, if transition probabilities are neglected and only densities of states of the involved bands are taken into consideration a broadening of the PL spectra with increaseing excitation energies should take place (observed by us recently in $GeSe_2$ glass (to be published)).

3.3 Luminescence intensity

It is generally observed that the PL intensity of a-SCs is weaker than that of the crystalline modifications. PL intensity in a-SCs varies linearly with excitation light intensity over a wide intensity range (see e.g. Figs 10 and 11) showing that PL is a result of monomolecular process(es). As a resultant of radiative and non-radiative processes the temperature dependece of PL intensity is strongly influenced by the respective transition probabilities p_r and p_{nr}, the former being independent of T, whilst the probability of non-radiative transitions increases exponentially with temperature:

$$p_{nr} \propto \exp(T) \tag{4}$$

As a consequence

$$\log\frac{p_{nr}}{p_r} \propto \log\frac{1-I_{PL}}{I_{PL}} \tag{5}$$

plotted as a function of temperature should give straight lines [24]. At very low temperatures in the range

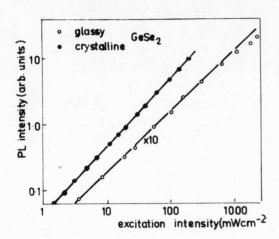

Fig. 10. Variation of PL intensity with excitation light intensity in crystalline and glassy GeSe$_2$ [23]

4.2-40 K some authors found that PL intensity did not depend on temperature (e.g. [25,26]); in ref. [24] it was a linear function of T in the range 4.2-23 K.

Scores of experimental data on PL intensity dependence on different parameters can be summed up in a simple phrase: sharp edges are favourable for PL. Accordingly, PL in a-SCs is much less intense then in c-SCs. Among amorphous SCs those with a stoichiometric composition will have the most intense PL: annealing and slow deposition rate would enhance PL. The PL intensity of bulk samples is

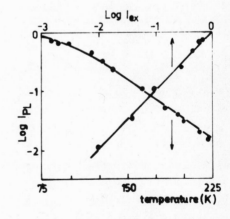

Fig. 11. The dependence of integrated PL intensity of Ge$_2$S$_3$ glass on the excitation light intensity and temperature [18]

usually greater than that of thin films, which is due partly to their
structure being disordered more extensively and partly to the increased
effect of surface states.

In a-As$_2$Se$_3$ a fast decay PL is observed with intensities much
weaker than the intensity of slow PL ($\tau_{PL} \sim 10^{-6}$-10^{-3} s). According to
ref. [20] slow decay PL is produced by distant electron-hole pairs,
whereas during the fast PL the charge carriers recombine at their place
of formation.

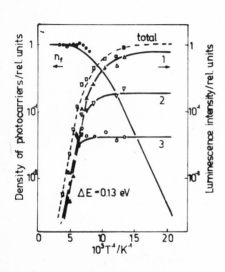

In connection with PL in-
tensity temperature dependence, an
important finding in a-Si [26,27]
is that PL intensity varies
strongly in a temperature range
when Δn, the concentration of the
excess free carriers, is constant
– as evidenced by photoconductivity
measurements (Fig. 12). Since no
change in the excitation and conduction process can be envisaged,
this means that rather than the
free carriers, electrically neutral
excited species are responsible
for PL. It is also of considerable importance in unravelling the
mechanism that PL intensity can be
decreased by applied electric
fields with a magnitude of about
10^5 volt/cm [26].

Fig. 12. Temperature dependence of the
density of photocarriers [28]
and of luminescence intensity
[26,27] of a-Si. The numbers
1, 2, 3 refer to three luminescence bands discussed in
section 2.4 of ref. [4].

3.4 PL decay kinetics

Studies of decay kinetics indicate that no unique time constant
exists which would characterize the whole life-span of PL: a fast PL
with τ = 10 ns was observed in As$_2$S$_3$ when excited by 5 nsec light
pulses [20], it takes values between 10 μs and 1 millisec when excited
with 5 μs duration pulses [29], and measured by a photodiode of
$\sim 2\times10^{-7}$ s time resolution. Ivashchenko et al. find the decay curves to
be composed of exponentials (Fig. 13), whilst Street's data [30] fit
neither first order nor second order kinetics. Preliminary irradiation
changes the decay kinetics as can be seen in Fig. 14. The first order
kinetics finding in ref. [29] is corroborated by the observations that
τ did not change when the excitation intensity was varied by two

orders of magnitude [18,23].

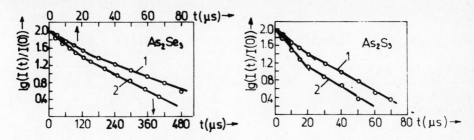

<u>Fig. 13.</u>   The decay of PL intensity at E_{PL}^{max} at 10 K (a)
            amorphous (1) and crystalline (2) As_2Se_3.
            (b) amorphous (1) and crystalline (2)
            As_2S_3 [29] .

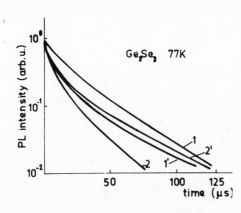

<u>Fig. 14.</u>   The decay of PL intensity
            at E_1=0.97 eV and E_2=0.87 eV
            for Ge_2Se_3. 1 and 2 non-ir-
            radiated, 1' and 2' irra-
            diated samples [18].

In a temperature range from 4.2 to 20-40 K (the upper limit depending on the amorphous materials), the decay time is constant, then it decreases exponentially with increasing temperature (Figs 15 and 16). This type of behaviour found in many cases can be rationalized by supposing that non-radiative transitions probably freeze-in at such low temperatures thus p_r gets close to unity and dominates the whole desexcitation.

Another distinguishing feature of PL decay in a-SCs is, at low temperatures, the dependence of γ on emitted energy in such a way that at longer wavelengths the decay is slower than at shorter ones [25,31]. This dependence becomes less pronounced at higher temperatures and disappeares in the temperature range where non-radiative desexcitation dominates the deactivation process.

3.5 Fatigue

Fatigue is the decay of PL intensity occurring *during* excitation. It is accompanied by the appearance of an induced absorption band in the forbidden gap due to species having unpaired spins. Highly absorbed light ($\alpha > 10^4$ cm^{-1}) causes weak PL which fatigues quite slowly, but

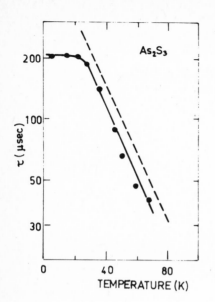

Fig. 15.

The luminescence decay time (defined as the slope of the decay curve after 100 μs), of a-As$_2$S$_3$ versus temperature. The dashed line represents the temperature dependence of the luminescence intensity [3,30].

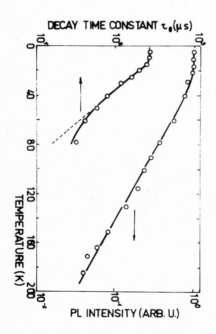

Fig. 16.

The luminescence decay time τ_0 and PL intensity as a function of temperature in GeSe$_2$ glass [25]

deeply penertrating light ($\alpha \leq 10^2$ cm^{-1}) induces intense PL which fatigues rapidly. This tendency observed on As$_2$Se$_3$ at 6 K [32] is reversed in the case of GeSe$_3$ at T = 2 K [33] (Fig. 17). These two examples - clearly not enough for widespread generalization - might indicate that there seems to be no simple connection between excitation

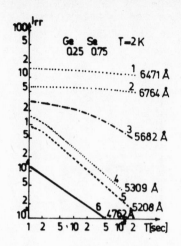

energy and fatigue rate.

At a given excitation energy the extent of the fatigue is proportional to the number of absorbed light quanta, i.e. a given percentage of fatigue needs a given number of photons to be absorbed by the material [32]. During fatigue at a given excitation energy PL intensity varies with excitation time as [3]

$$I \propto t^{-b} \qquad (6)$$

where t is the excitation time at a given energy and b is a constant depending on excitation light intensity and wavelength. It means that PL fatigue rate at a given excitation wavelength is a measure of the light intensity.

Fatigue does not usually lead to the complete disappearance of PL but instead its intensity reaches a steady state value depending on the light intensity and the composition of the material. The Ge-Se system serves to illustrate the effect of composition: the steady state PL intensity value is nearly zero in alloys of Ge_xSe_{1-x} when x is small; it is $\sim 1/2 \, I_o$ near x = 0.33; there is only a very weak fatigue in $c-GeSe_2$; and no fatigue has been observed in c-GeSe [33], while fatigue was not observed neither in glassy nor in crystalline $GeSe_2$ at 77 K [14].

At steady state, the rates of excitation and desexcitation are in equilibrium, moreover p_{nr}/p_r remains constant. During the fatigue of PL some of the radiative centres are converted to non-radiative ones and the concentration of the latter reaches a limiting value. If fatigue were connected with the population and depopulation of long lived preexisting traps of extrinsic or intrinsic origin, crystalline samples would not luminesce or would only weakly do so, and a steady-state PL value determined by impurity concentration both in amorphous and crystalline states would be observed. Although relevant data are not consistent, they can more easily be rationalized if one supposes that both the properties and the concentration of traps change during prolonged excitation.

There are interesting data indicating that fatiguing processes take place over a much shorter time scale than those usually registered (Fig. 18 and 19). In $GeSe_2$ a "fast" ($\tau \approx$ 5-20 s) fatigue was found [34]

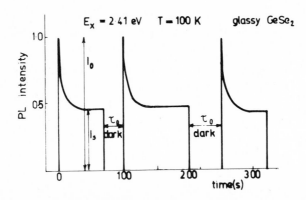

Fig. 18. The variations of PL intensity with time at 100 K during continuous excitation with light quanta of E_x = 2.41 eV energy

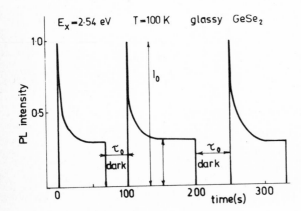

Fig. 19. The variations of PL intensity with time at 100 K during continuous excitation with light quanta of E_x = 2.54 eV energy.

whose rate depended on excitation energy. The intensity of PL corresponding to steady-state diminishes as the energy of exciting light quanta are increased. A peculiar feature of this fast fatigue is that the sample retains its initial properties if it is kept in the dark for a certain time, not less than τ_{min} after the steady-state has been reached and the excitation was stopped. For complete restoration to occur τ_{min} = 20-30 s is needed. These observations can be explained by supposing that in the initial stage of the slow fatigue some of the excited species are trapped in shallow traps for a few seconds, then subsequently get thermally liberated and recombine non-radiatively leading to the restoration of the initial trapping properties of the material.

3.6 PL quantum efficiency

Quantum efficiency of PL shows the ratio

$$\eta_{PL} = \frac{P_r}{P_{nr} + P_r}$$

during the desexcitation process. If we suppose - as Mott did [31] - that the non-radiative process is thermally activated, then

$$\eta_{PL} = \frac{1}{1 + (\omega/A)\exp(-W/kT)} \qquad (7)$$

where W is the activation energy, ω is the lattice frequency, $\omega/A = 10^4$ as suggested from independent measurements. Equation (7) predicts an increase in η_{PL} if W is large and kT is small, in agreement with experimental data.

There has been no direct measurement of η_{PL}, so indirect rough estimates are used instead. It is felt that the formation of ESR centres and PL fatigue are closely connected; the decrease in η_{PL} due to fatigue is accompanied by the accumulation of ESR centres. A rough estimate of P_r and η_{PL} can be obtained by supposing that the ESR centres' accumulation rate is of the same order as the radiative transitions rate. Following this line of thought one encounters gross discrepancies in explaining some of the experimental findings: e.g. in $GeSe_2$, N_s (the saturation value of the spin concentration) is $\sim 10^{15}$ cm^{-3} while PL is at least as efficient as in As_2Se_3 for which $N_s = 10^{17}$ cm^{-3}. It is to be noted that η_{PL} in As_2S_3 was estimated to be ~ 0.1-0.2 [3] and in $GeSe_2$ glass ~ 0.5-1.0, suggesting that photo induced ESR centres and those for radiative recombination might not be connected so strongly as proposed in [49].

Admittedly there is some connection between PL and ESR centre formation although these processes cannot be linked so closely to each other. Perhaps the most useful estimates on the nature of basic processes can be deduced from data characterizing the steady-state.

3.7 Induced absorption spectrum and ESR centres

Excitation with bandgap irradiation does not only produce PL but also leads to the appearance of an absorption spectrum starting at Eg/2 and overlapping with the absorption tail [2,11,35]. This induced absorption band is due to long lived trapped charge carriers. This band and the ESR signal are due to the same species, owing to the fact that both the induced absorption band and the ESR signal can be eliminated by heating the sample or by irradiating it with IR light of energy corresponding to the absorption band associated with them.

The absorption coefficient α equals 30 cm^{-1} for the induced band
in a-Si [32]. Taking this value and a supposed cross section of 10^{-16}
cm^2 within the range 0.8-1.55 eV yields the concentration of these
centres being in metastable states as $N_s = 10^{17}$ cm^{-3}. Roughly the same
value is arrived at from ESR measurements.

A remarkable fact having a great significance when models are
elaborated is that the concentration of ESR active centres and probably
the induced spectrum (the latter not measured yet) during prolonged
excitation reach a steady-state value, the time for reaching it being
dependent on the excitation light intensity. It is interesting that in
glasses of the system Ge-Se ESR centres do not saturate during irradiation, while PL intensity reaches a steady-state value in the fatiguing
process.

4. MODELS

The dominant feature of PL in amorphous chalcogenide semiconductors is the large Stokes shift. Thermalization in band tails [36] , and
the formation of neutral D^o and charged D^- and D^+ centres [37] have
been suggested as a way of rationalizing experimental data. The
models based on these and similar concepts have been successful in explaining most of the experimental findings but they have recently come
up against obstacles in attempting to accomodate the latest results,
especially those obtained by ESR and photoconductivity techniques.

4.1 The Mott-Street model

Basic to the Mott-Street (MS) model (see e.g. [37]) is the supposition of defects intrinsically connected with the bonding properties
of the chalcogen atoms. Being intrinsic, these structural defects
appear both in the crystalline and in the amorphous forms of the materials.
Dangling bonds in the bulk are thought as dominating the electrical
and optical properties of amorphous chalcogenide semiconductors. The
principal feature of this model is the proposal that the neutral
dangling bond D^o is not stable, but a pair of them would ionize

$$2 \ D^o \longrightarrow D^+ + D^- \tag{8}$$

and pass to stable charged centres, localized near the v- and c-band
edges. Accordingly, the lowest energy absorption is the promotion of
an electron from the D^- state, situated near the v-band to the conduction band or alternatively transferring a hole from a D^+ state to
the valence band. It was later shown by Kastner et al. [38,39] that

the empty or doubly occupied dangling bonds in chalcogenides correspond
to a three-fold coordinated C_3^+ positive and a singly-coordinated C_1^-
negative ion. They, too, retained the basic supposition that their
formation lowers the energy of the system since

$$2\ C_2^o \longrightarrow C_3^+ + C_1^- \tag{9}$$

is an exothermic process.

In this model the excitation process creates $D^o(C_1^o)$ centres lying
deep in the gap, which might recombine radiatively with the other
carrier. If instead this carrier tunnels away, it leaves behind an
isolated metastable paramagnetic centre. Such isolated centres are
responsible for the fatigue of PL, induced optical absorption and ESR
signals.

The main argument in favour of the involvement of charged centres
in the excitation process is that PL efficiency falls off very rapidly
with increasing temperature [31]. This, as interpreted by Street [3],
shows that after ionization the centre is neutral since in the opposite
case a Coulomb field would exist and the temperature effect would be
much less pronounced. A thorough examination of the existing results
has demonstrated weak points in this model. Some of the findings not
easily accomodated in it are as follows:

- The MS model considers only one type of defect giving one type
of centre with unpaired spin D^o, whereas according to the observed ESR
spectra in As-containing chalcogenides there exist two types of optically induced centres, viz. an electron missing from a non-bonding
p-orbital, and an electron localized on an As atom [2].

- According to Kastner [39] the most likely metastable optically
induced paramagnetic neutral defect (replacing D^o in the MS model) is
C_3^o, an unpaired electron in the antibonding state; the most probable
centre found by ESR is a hole.

- Four coordinated pnictide atoms (P_4^o) proposed as D^o for VAPs in
$A^V B^{VI}$ chalcogenide glasses are not consistent with the ESR findings:
P_4^o $(s^2 p^3)$ would require extensive s p hybridization not confirmed by
the ESR singal, which is best explained by an unpaired spin in a
single p-orbital with only a small (5 %) s-orbital admixture.

- The MS model predicts nearly identical radiative centres for
the crystalline and amorphous forms of chalcogenides. Indeed PL spectra
in crystalline and glassy As_2Se_3 and As_2S_3 are very similar, but
fatigue of PL and ESR active centres can be found only in the glassy
compounds.

- In this model the fatigue of PL is accompanied by the formation of D^O, a neutral particle situated at the middle of the gap, which does not account for the increase in optical absorption in the absorption tail - an observed fact (see e.g. [16]).

- In GeSe$_2$ the maximum PL and ESR efficiency appear at different excitation energies: PL reaches its maximum value at a lower energy. In addition, it is established that there are two kinds of ESR signal in the Ge$_x$Se$_{1-x}$ if $x < 0.2$ and their relative concentration depends on E_x; red light created centres reach a saturation value of $N_s = 3 \times 10^{17}$ cm^{-3} and are connected with radiative transitions, whereas those formed by blue light are more abundant ($N_s = 10^{20}$ cm^{-3}) and are probably non-radiative [40].

-- It is difficult to interpret PL fatigue data obtained in the system Ge$_x$Se$_{1-x}$: PL intensity, in contrast with other compositions, does not decrease in GeSe$_2$ glass.

4.2 Model based on Onsager's theory

At present, the most natural model for describing PL in chalcogenide glasses is based on Onsager's theory on the random walk dissociation and recombination of photo-excited electron hole pairs [41,42, 56,56]. As envisaged by this model, on photon absorption an electron is promoted from the valence band to the conduction band. If the incident photons are of higher energy than E_g then the newly created excess electrons and holes are formed relatively deep in the respective bands and will have an extra kinetic energy of $h\nu - E_g$ above the band edge. This extra kinetic energy will be dissipated by emission of phonons.

Assuming that the movement of the excited charge carriers during the thermalization process is by diffusion the thermalization distance is given by $r_o = (D\tau)^{1/2}$ (where D is the diffusion coefficient and τ is the thermalization time). Defining a critical distance r_c where the Coulomb potential energy $e^2/\varepsilon r_c$ becomes equal to the average thermal energy kT, it can be easily seen that an electron thermalized at distance r_o from its parent hole will either form an exciton and move together or they diffuse apart from each other depending on the ratio r_c/r_o. The escape probability is $\exp(-r_c/r)$ where r_c, the Coulombic capture radius from the above mentioned criterion, is $r_c \equiv e^2/\varepsilon kT$. It can be seen that the escape probability for electrons thermalized within the radius of r_c is small. By lowering the temperature the Coulomb capture radius r_c will increase. In materials with low ε, at low temperature it can reach hundreds of nearest neighbour distance.

The total thermalization time can be given as

$$\tau = \beta \ \frac{(h\nu - E_g) + \dfrac{e^2}{\varepsilon r_o}}{h\,\bar{\nu}_{ph}^{\,2}} \tag{10}$$

where β is a coefficient characteristic of the electron-phonon coupling, $\bar{\nu}_{ph}$ is the typical phonon frequency and $e^2/\varepsilon r_o$ is the binding energy of the electron-hole pair at r_o separation. For r_o the following equation is obtained:

$$\frac{r_o^2}{D} = \beta \ \frac{(h\nu - E_g) + \dfrac{e^2}{\varepsilon r_o}}{h\,\bar{\nu}_{ph}^{\,2}} \tag{11}$$

From photoexcitation experiments of a-Se [57] r_o values of 70 and 12 Å were deduced at excitation wavelengths of 400 and 580 nm respectively.

Mobilities in chalcogenide glasses are low, even trap-free mobilities are thought to be less than ~ 1 cm^2/voltsec, thus the charge carriers move slowly in these materials and spend a comparatively long time at a given site between hops. During residence time slow electrons (holes) might polarize their surroundings and lower their energy by occupying a bound state in the self-induced potential hollow. The charge carriers dig their own well and get self-trapped. Self-trapping may occur at different distances from the nearest oppositely charged carrier, thus they would feel each other's coulombic field differently (according to the distance separating them).

In chalcogenides the electron-hole pairs are normally not separated except at high enough temperature and in the presence of an applied electric field [58], thus the absorbed photons create thermalized electron-hole pairs. (For comparison we note here, that in amorphous Se each absorbed photon creates a thermalized electron-hole pair [57], why in anthracene and poly(N-vinil-carbazol) this ratio is 10^{-3}-10^{-4} [59] and ~ 0.25, respectively [60]. In other words, excited electrons thermalize and get self-trapped close to their parent holes, so that the Coulomb attraction between them is strong.)

The situation is depicted in Fig. 20. Self-trapped charge carriers feeling each other's coulombic field can recombine by tunnelling or after being thermally freed. Such recombination occurs directly and no recombination centres like preexisting defects, impurities, or dangling bonds need to be involved.

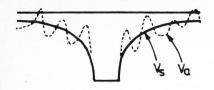

Fig. 20. Energy scheme of a Coulomb centre with traps [43]

Those pairs which recombine radiatively emit photons with energies much lower than that of the exciting radiation, since one or both of the carriers underwent a decrease in energy as a consequence of self-trapping. Thus the large Stokes shift finds a natural explanation. The situation can be depicted by a configuration coordinate diagram as shown in Fig. 21.

During prolonged excitation the number of pairs which became thermalized beyond r_c increases. They get self-trapped and are unable to move without the necessary activation energy, so their recombination is inhibited. With the increase in the concentration of such metastably separated pairs the quantum efficiency of PL will drop. This process is the so called fatigue, which is accompanied by the appearance of an induced absorption band and ESR active centres - both of which are due to the self trapped holes and/or electrons.

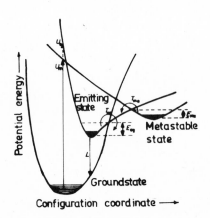

Fig. 21. A possible configurational coordinate diagram for PL in chalcogenide glasses [44].

ESR measurements show that the concentration of these metastable species reaches a saturation (steady-state) value. On reaching this concentration, oppositely charged carriers probably get so close to each other that their wavefunctions overlap substantially and they may recombine by tunnelling. The distance r_s separating the charges, after the steady-state had been reached can be figured out from the steady-state concentration of ESR active centres, since

$$N_S = (4 \pi r_s^3/3)^{-1} \qquad (12)$$

where N_S is the saturation concentration of the uncompensated spins. If $N_S = 10^{18}$ cm^{-3}, a value commonly found in chalcogenides, $r_s \sim 60$ Å, where the Coulomb attraction is 0.24 meV.

Even though this classical picture of thermalization seems to be quite straightforward, on closer examination it reveals many weak points [43]. A quantum mechanical description of thermalization shows that in the condensed phase no principal differences exist between excitons and separated electron-hole pairs. Because of the wave nature of the electron the thermalization process has some peculiarities, namely, that the electron motion is attenuated with increasing electron-scattering cross section when the excitation energy either increases or decreases relative to the ionization potential, a parameter not precisely determinable in systems with large fluctuating potentials acting as traps. It is also noteworthy that from the uncertainty principle $\Delta r = \hbar v / kT$, where v is the charge velocity. That is: at $T = 300$ K the uncertainty in the position of the electron is about 20 Å. The change in Coulomb potential within 20 Å might be many times that of kT.

5. OUTLOOK

Significant changes in the Mott, Street, Davis and the Kastner, Adler, Fritzsche models should be made when attempting to reflect on the kinetical data accumulated so far. Maruyama's fast decay kinetics [20] and our fast fatigue measurements [34] show that decay and fatigue mechanisms change considerably with time; it is not the same throughout the whole process but rather can be divided into separate ranges characterized by quite different elementary processes. The fact that the initial fatigue caused by ~ 20 s illumination is not permanent makes a trapping mechanism as its cause more probable than the electron escape proposed in the MS model. More generally, it can be expected that pulse measurements of PL spectra, PL decay and fatigue will reveal new features of these processes, e.g. very fast decaying band gap PL, and make possible to follow spontaneous changes of such parameters as trap depths, absorption edges, etc.

The connection of fatigue with pre-existing traps seems to be corroborated by the lack of or only a very weak fatigue in crystalline counterparts of the chalcogenide glasses. The more ordered structure of the crystals is initially free of trapping sites and the weak fatigue sometimes observed is probably due to traps formed as a consequence of prolonged irradiation with light.

The lack of any effect of comparatively strong electric fields $(E \sim 10^5$ V/cm) on the characteristic of PL in $GeSe_2$ [45] and As_2Se_3 [46] speaks of a very high field between the electron and hole in the pairs both pre-existing and created by photon absorption. In other

words, sibling electrons and holes are very close to each other; ten
angstroms or a few times that value is a probable distance separating
them.

The description based on Onsager's model can accomodate both pre-
existing and photo-induced self-trapped electron-hole pairs and treat
them equally as potential radiative and non-radiative centres. We suggest that PL is a consequence of charge transition within these pairs.
Fatigue is caused by the self trapping either of electrons or holes or
both. The traps thus formed are initially shallow, this means that
thermal detrapping is possible, but later they spontaneously relax and
deepen to such an extent that detrapping is thermally no longer
feasible. The deepening of traps leads to a metastable state of the
excited system. The process of trapped charge accumulation appears as
fatigue of PL, since the charges trapped in pre-existing, originally
neutral traps cannot recombine unless they had been freed by IR irradiation or by having sufficiently heated up the fatigued sample.

If one accepts this model, the chalcogenid glasses could be united
and commonly treated with other glasses - both inorganic and organic -
in which the origin of the induced optical properties and ESR signals
are quite well understood [47,48]. In this framework chalcogenides
remind one of the behaviour of mildly polar organic glasses characterized by moderate dielectric coefficients. The PL of chalcogenides
might soon fit into the general picture of F centres, self-trapped
electrons and impurity colour centres.

ACNOWLEDGEMENTS - The authors would like to acknowledge helpful discussions with Drs B.L.Gelmont and P.Fazekas.

REFERENCES
1. B.T.Kolomiets, T.N.Mamontova and V.V.Negreskul, Phys.Stat.Solidi
    27, K15 (1968)
2. S.G.Bishop, U.Strom and P.C.Taylor, Proc. 7th Int. Conf. Amorphous
    and Liquid Semiconductors, ed. by W.E.Spear, Edinburgh 1977,p.595
3. R.A.Street, Advances in Physics 25, 397 (1976)
4. R.Fischer, Feskörperprobleme. XVII. Advances in Solid State
    Physics p.85, ed. by J.Treusch, Vieweg, Braunschweig, 1977.
5. N.F.Mott, Advances in Physics 16, 49 (1967)
6. M.-L.Theye, Rev.Physique Appl. 12, 725 (1977)
7. M.Lanno, M.Bensoussan, Phys. Rev. B 26, 3546 (1977)
8. M.Bensoussan, Rev. Physique Appl. 12, 753 (1977)
9. T.T.Nang, M.Okuda, T.Matsushita, S.Yokota and A.Suzuki, Jap.J.
    Appl.Phys. 15, 849 (1976)
10. N.Ch.Abrikosov and L.E.Shelimova, Semiconducting materials based
    on $A^{IV}B^{VI}$ compounds. Nauka Publ. Moscow 1975, pp.7-13 (in Russian)
11. J.Shirafuji, G.I.Kim and Y.Inuishi, Jap.J.Appl.Phys. 16, 67 (1977)
12. M.Kumeda, M.Ishikawa, M.Suziki and T.Shimizu, Solid State Comm.
    25, 933 (1978)

13. R.A.Street, I.G.Austin and T.M.Searle, J.Phys. C8, 1293 (1975)
14. V.A.Vassilyev, M.Koós and I.Kósa Somogyi, Solid State Commun. 22, 633 (1977)
15. R.A.Street, T.M.Searle and I.G.Austin, Phil.Mag. 29, 1157 (1974)
16. V.A.Vassilyev, S.K.Pavlov and B.T.Kolomiets, Amorphous Semiconductors'76, Proc. Int. Conf. ed. by I.Kósa Somogyi, Akadémiai Kiadó, Budapest, 1977, p.189
17. W.Henrion and M.Zavetova, Proc. Conf. "Amorphous Semiconductors'74" Reinhardsbrunn, GDR, 1974, p.280
18. B.T.Kolomiets, T.N.Mamontova and V.A.Vassilyev, Structure and Properties of Non-Crystalline Semiconductors, ed. by B.T.Kolomiets Nauka, Leningrad, 1976, p.227
19. L.Tóth, V.A.Vassilyev and I.Kósa Somogyi, Report, KFKI-1978-27
20. M.Maruyama, T.Ninomiya, H.Suzuki and K.Morigaki, Solid State Comm. 24, 197 (1977)
21. B.T.Kolomiets, T.N.Mamontova and A.A.Babaev, J.Non-Crystalline Solids 4, 289 (1970)
22. B.T.Kolomiets, T.N.Mamontova, A.A.Babaev, J.Non-Crystalline Solids 8-10, 1004 (1972)
23. M.Koós, I.Kósa Somogyi and V.A.Vassilyev, Proc. Int. Conf. Amorphous Semiconductors'78 (to be published), Pardubice, 1978.
24. I.S.Shlimak and R.Rentzsch, Amorphous Semiconductors'76, Proc. Int. Conf. ed. by I.Kósa Somogyi, Akadémiai Kiadó, Budapest 1977, p.177
25. V.A.Vassilyev, M.Koós and I.Kósa Somogyi, Phil.Mag. (to be published)
26. D.Engemann and R.Fischer, Structure and Excitations in Amorphous Solids, AIP Conf. Proc. N°31, American Institute of Physics, New York 1976, p.37
27. D.Engemann and R.Fisher, phys.stat.sol.(b) 79, 195 (1977)
28. W.E.Spear, R.J.Loveland and A.Al-Sharbaty, J.Non-Crystalline Solids 15, 410 (1974)
29. Yu.N.Ivashchenko, B.T.Kolomiets, T.N.Mamontova and E.A.Smorgonskaya, phys.stat.sol.(a) 20, 429 (1973)
30. R.A.Street, T.M.Searle and I.G.Austin, Amorphous and Liquid Semiconductors, ed. J.Stuke and W.Brenig (Taylor and Francis) London 1974, p.953
31. N.F.Mott, Phil.Mag. 36, 413 (1977)
32. S.G.Bishop and U.Strom, Optical Properties of Highly Transparent Solids. Ed. S.S.Mitra and B.Bendow, Plenum N.Y. and London 1975. p.317
33. J.Cernogora, F.Mollot, C. Benoit á la Guillaume and M.Bensoussan, Amorphous and Liquid Semiconductors, ed. W.A.Spear, Proc. 7th International Conf. Edinburg, 1977. p.617
34. V.A.Vassilyev, M.Koós and I.Kósa Somogyi, Solid State Commun. 28, 634 (1978)
35. S.G.Bishop, U.Strom and P.C.Taylor, Structure and Excitation of Amorphous Solids, eds. G.Lucovsky and F.L.Galeener, AIP Conf. N°31, N.Y. 1976, p.16
36. R.Fischer, U.Heim, F.Stern and K.Weiser, Phys.Rev.Lett. 26, 1182 (1971)
37. R.A.Street and N.F.Mott, Phys.Rev.Lett. 35, 1293 (1975)
38. M.Kastner, D.Adler and H.Fritzsche, Phys.Rev.Lett. 37, 1504 (1976)
39. M.Kastner, Amorphous and Liquid Semiconductors, ed. W.E.Spear, Edinburgh, 1977, p.504
40. C. Benoit á la Guillaume, F.Mollot and J.Cernogora, Amorphous and Liquid Semiconductors, ed. W.E.Spear, Edinburgh, 1977, p.612
41. L.Onsager, J.Chem.Phys. 2, 599 (1934)
42. L.Onsager, Phys.Rev. 54, 554 (1938)

43. K.Funabashi, Excess electron processes in radiation chemistry of disordered materials. Advances in Radiation Chemistry, Vol.4. ed. M.Burton and J.L.Magee, John Wiley and Sons, N.Y. 1974, pp. 103-180
44. P.Kivits, M.Wijnakker, J.Claassen and J.Geerts, J.Phys.C: Solid State Phys. 11, 2351 (1978)
45. M.Koós and I.Kósa Somogyi (unpublished results)
46. S.J.Hudgens and M.Kastner, in Amorphous and Liquid Semiconductors, ed. W.E.Spear, Proc. 7th Int. Conf. on Amorphous and Liquid Semiconductors, Edinburg, 1977, p.622
47. J.E.Willard, J.Phys.Chem. 79, 2966 (1975)
48. L.Kevan, Advances in Radiation Chemistry, ed. by M.Burton and J.L.Magee, Wiley-Interscience N.Y. 1974, pp.275-298
49. S.G.Bishop, U.Strom and P.C.Taylor, Phys.Rev.Lett. 34, 1346 (1975)
50. R.A.Street, Phys.Rev. B17, 3984 (1978)
51. T.T.Nang, M.Okuda and T.Matsushita, J.Non-Crystalline Solids 33, 311 (1979)
52. J.Shah and M.A.Bösch, Phys.Rev.Letters 42, 1420 (1979)
53. V.A.Vassilyev, T.N.Mamontova and B.T.Kolomiets, Amorphous Semiconductors'76, ed. I.Kósa Somogyi, Akadémiai Kiadó, Budapest, 1977, p.178
54. G.Lucovsky, Phys.Rev. B15, 5672 (1977)
55. E.A.Davis, J.Non-Crystalline Solids 4, 107 (1970)
56. J.E.Knights and E.A.Davis, J.Phys.Chem.Solids 35, 543 (1974)
57. D.M.Pai and R.C.Enck, Phys.Rev. B11, 5163 (1975)
58. N.F.Mott, Solid State Electronics 21, 1275 (1978)
59. R.M.Batt, C.L.Braun and J.F.Horning, J.Chem.Phys. 49, 1967 (1968)
60. P.J.Melz, J.Chem.Phys. 57, 1964 (1972)

MAN-MADE SEMICONDUCTOR SUPERLATTICES

G. A. Sai-Halasz
IBM Thomas J. Watson Research Center
Yorktown Heights, N. Y. 10598 USA

ABSTRACT

A review is presented on man-made $GaAs-Ga_{1-x}Al_xAs$ and $In_{1-x}Ga_xAs-GaSb_{1-y}As_y$ heterostructures. Emphasis is given to experimental work and to the recent developments in the field.

When in a crystal in one dimension, say z direction, a periodic potential, $V(z+d)=V(z)$, is superinposed on the basic lattice periodicity, such that d is microscopic but much larger than the lattice period, a superlattice (SL) structure is realized. Usually the superpotential can be regarded as a perturbation, to be delt with in the context of the underlying band-structure of the host crystal. Naturally occurring SL-s are frequent in various materials, but only in the past 7-8 years have they been produced artificially in semiconductors.

Interest in semiconductor SL-s arose when Esaki and Tsu [1] showed that SL-s can exhibit negativ differential resistance and Bloch oscillations. The same authors also proposed realistic ways for their fabrication. More detailed theoretical treatments indicated that SL frequency capabilities may indeed be in the THz range [2-6]. The noteworthy properties of SL-s are due to the subband structure which is a consequence of the superpotential. The Brillouin zone shrinks in the z direction and there will be a series of bands, called the subband structure, as indicated in Fig. 1a for the case of a rectangular superpotential. The dispersion in the z direction, illustrated in Fig. 1b, depends on the strength and period of the superpotential. To first approximation, in the x-y plane the dispersion of the host remains unperturbed. The resulting density of states is shown in Fig. 1c.

Considerable effort was expended on the investigation of SL-s in the past 7-8 years. Primary motivation was that one had at hand a system in which simple but fundamental experiments could be performed. Fundamental physical systems and effects, like particle in a box, coupling between wells, interaction between states of different parity, variable dimensionality etc. became realizeable. Also, the possibility of creating new devices provided additional incentives.

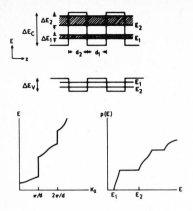

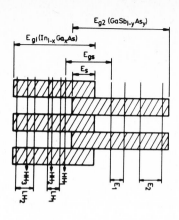

Fig. 1.   a) Potential energy profile of
a Type I SL. b) Dispersion in
the z direction. c) Density of
states due to the subband
structure

Fig. 2.   Potential energy profile of a
Type II SL. E_s is the band-edge
overlap parameter, E_{gs} is the SL
gap.

Of the various proposals for realizing a semiconductor SL so far
only the heterostructure proved successful. In a heterostructure two
semiconductors are periodically deposited in ultrathin layers. The potential energy discontinuities at the interfaces provide the superpotential. To observe LS effects one requires very high quality materials
and atomically smooth interfaces. Vapor phase deposition yields high
quality materials and interfaces [7], but to date the shortest period
grown by this method is about 180 $\mathring{A}$ [8]. Molecular Beam Epitaxy (MBE)
has been the workhorse of the SL research, and vice versa, the severe
materials quality requirements posed by the LS-s stimulated the
progress of the MBE technique.

Most of the work has been done on the GaAs-Ga$_{1-x}$Al$_x$As system.
GaAs and Ga$_{1-x}$Al$_x$As besides having closely matched lattice constants
in the heterostructure form simple superpotential, since the GaAs
valence band (VB) edge is at higher energy and its conduction band
(CB) edge is at lower energy than the corresponding edges in Ga$_{1-x}$Al$_x$As.
Thus, the Ga$_{1-x}$Al$_x$As represents a potential barrier for the electrons
in the vicinity of both band edges, as illustrated in Fig. 1a. We call
a SL with this band-edge configuration Type I. SL. An other band-edge
configuration is shown in Fig. 2. Here the CB and VB edges of one semiconductor lie above the corresponding edges of the other. A SL with
InAs and GaSb, or their alloys with GaAs, as hosts represent such a
system [9]. These are called Type II. SL-s. The energy overlap para-

meter, E_s, also shown in the figure, is defined as the energy difference between the CB edge of $In_{1-x}Ga_xAs$ and the VB edge of $GaSb_{1-y}As_y$.

Historically, the first experiments indicating superlattice formation were done in MBE grown $GaAs-Ga_{1-x}Al_xAs$ SL-s. Negative differential resistance was observed in a narrow period SL [10]. In SL-s with wide $Ga_{1-x}Al_xAs$ barrier layers, resulting in narrow subbands, oscillatory conductivity was observed [11]. Results have been interpreted on the basis of quantum levels in successive wells switching into resonant condition, as shown in Fig. 3. Subsequent optical absorption in single [12] and multiple [13] GaAs wells showed clearly the energy positions of the quantum levels, and how the coupling of these levels leads to formation of subbands. At the same time, one accurately could deduce the potential discontinuity between GaAs and $Ga_{1-x}Al_xAs$ [12]. These early experiments showed that the simple picture, of one-dimensional wells and barriers with electrons of appropriate effective masses, worked remarkably well. The calculated subband structure agreed with the experimentally measured one to within a few percent. Next, the optical and transport behaviour of the SL-s was correlated in photoconductivity measurements [14]. Fig. 4a shows that for a given applied voltage the photocurrent peaked at photon energies coinciding with allowed subband transitions. With the photon energy tuned at one of these peaks, the photocurrent exhibited negative differential resistance, see Fig. 4b.

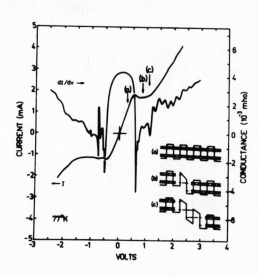

Fig. 3. I-V characteristics of a tight binding SL.

Dimensionality effects and their variation with SL parameters were studied in resonant Raman scattering [15] (RRS) and Shubnikov-de Haas [16] (S-dH) measurements. A two-dimensional electronic system, under resonant conditions and in the absence of lifetime broadening, gives rise to singularities in the Raman scattering cross section. However, the effect has not been observed experimentally because two-dimensional systems generally have insignificant volume, which leads to undetectable scattering. On the other hand, in a SL one can have

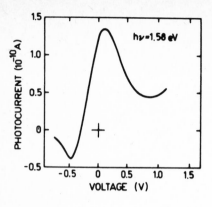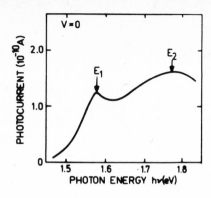

Fig. 4. a) Photocurrent vs. photon energy for a SL. Calculated subband transition
energies are indicated with arrows. b) I-V curve of the photocurrent

the combined scattering of hundreds of identical wells, and, indeed,
greatly enhanced scattering was observed in SL-s in the "forbidden"
geometry. The relative intensities were related to the degree of two-
-dimensionality of the subbands. Fig. 5 shows the agreement between
the measured and calculated enhancements in different samples. RRS
also permitted the demonstration [17] that the fundamental periodicity
in a SL is d: umklapp processes with multiples of $2\pi/d$ wavevectors
were observed. An other interesting feature in these experiments was
the indication of subband formation in the spin-orbit split VB [18].
S-dH oscillations in the current flowing parallel to the layers showed
how the degree of two-dimensionality affects magneto-transport. The
theory which took into account subbands was in excellent agreement
with the observed oscillatory conductivity, but so far has not explained
the pronouced negative magnetoresistance characteristics of the SL-s
[19]. Independently, magneto-absorption measurements were also interpreted successfully in the context of subband structure [20]. The MBE
technique made it possible to achieve close to monolayer structures of
GaAs and AlAs [21]. These structures in the strictest sense are not
superlattices but rather new kind of materials. At this point, however,
it is not clear whether the monolayers differ significantly from a
random $Ga_{.5}Al_{.5}As$ alloy.

The latest development in the $GaAs-Ga_{1-x}Al_xAs$ SL-s is the realization of the so called modulation doping technique [22]. Here one
introduces donors exclusively in the $Ga_{1-x}Al_xAs$ layers and the electrons

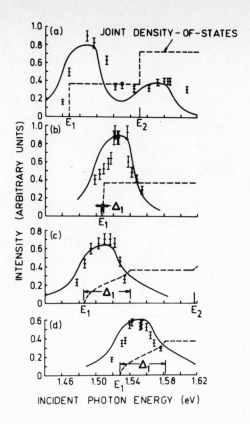

Fig. 5.   Raman intensity vs. incident-photon energy for four SL configurations,
solid curves are calculated. The energy dependence of joint density of
states is shown in dashed lines.

drop into the GaAs layers forming a two-dimensional electron gas without the presence of the parent impurities. Due to lack of impurity
scattering in the GaAs layers, mobilities as high as 20,000 cm^2/Vsec
were measured at carrier concentration of $5x10^{16}$/cm^3 at helium temperatures.

The interesting transport features of SL-s arise from the negative
effective mass region of the subbands. Properties which involve both
the CB and VB depend on the spacial realtion of the envelope wavefunctions in these two bands. Heterostructures made of InAs and GaSb
and their alloys with GaAs can have increased negative mass regions in
the Brillouin zone, and always exhibit Type II. band-edge lineup at
the heterojunction interfaces. In the energy range near the bottom of
the InAs CB the InAs and GaSb heterojunction represents an "up side down"

potential configuration. An "up side down" potential means that the
barriers become more difficult to penetrate as the energy of the tunneling electron increase [9]. Under these conditions the inflection
point in the subzone dispersion will move toward the zone center. In
Fig. 6 the gaps of $In_{1-x}Ga_xAs$ and $GaSb_{1-y}As_y$ are plotted against the
GaAs content, as obtained
from electron affinity [23]
with the zero of the energy
scale at the vacuum level.
The electron affinity measurements coupled with
theoretical considerations
[24], and heterojunction
I-V investigations [25] indicate that there is direct
contact between the VB of
GaSb and the CB of InAs,
i.e. a negative E_s as defined in Fig. 2. Different
considerations, on the other
hand, predict that charge
rearrangement alter the interface such that E_s between

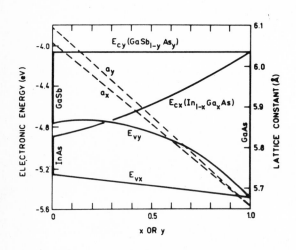

Fig. 6. The changes of band-edge energies and
lattice constants with compositions x and
y in the $In_{1-x}Ga_xAs-GaSb_{1-y}As_y$ system.

InAs and GaSb would become significantly positive [26].

Keeping the original one dimensional treatment, which was successful for Type I. SL-s, and taking just one step toward complexity by
introducing Bloch function into the description of the hosts, one ob-
taines the subband structure for the InAs-GaSb SL [27], as shown in
the middle of Fig. 7. Even for the assumed negative E_s there is a well
defined gap, meaning that the InAs-GaSB SL is a semiconductor for the
chosen layer thicknesses. Also, the first CB subzone has a wide region
where the effective mass is negative, as expected. When E_s is negative
the SL gap is strictly due to confinement [27]. Consequently, when the
period increases and confinement becomes less important the gap has to
go to zero. According to our calculation, this semiconductor to semimetal transition should occur when the layers are ~ 100 Å thick.
Fig. 8 shows how the neighbourhood of the band edges appear [27] in
thick SL-s when electrons spill from the GaSb VB into the InAs CB.

$GaSb_{1-y}As_y$ and $In_{1-x}Ga_xAs$ have been successfully grown by MBE
[28], and on the basis of S-dH [29] and optical absorption [30] measurements it is evident that we are able to fabricate high-quality

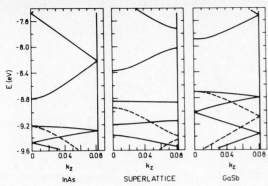

Fig. 7.

Band structure for SL-s in
the vicinity of their
fundamental gap, each 12
atomic planes thick. On the
side panels both host layers
are of the same material,
shown for comparison with
the InAs-GaSb SL. The light
hole bands are shown as
dashed lines.

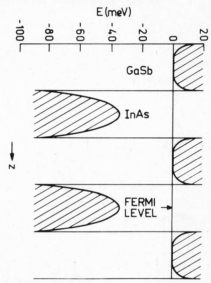

SL-s from these alloys. The
optical measurements also
clearly showed that E_s is
indeed negative between
InAs and GaSb.

The S-dH measurements
indicated two-dimensional
subbands in heavily n-doped
pure InAs-GaSb samples with
periods between 110 and
400 Å. Fig. 9a shows that
with decreasing angles between the applied magnetic
field and the plane of the
layers the oscillations in
the parallel conductivity
shift toward higher field.

Fig. 8. Band edges and Fermi level for a
500 Å - 500 Å InAs-GaSb SL. The shaded
regions are the gaps of the hosts.

Analysis, as in Figs. 9b and c, showed that this feature is due to the
two-dimensional nature of the electronic states. In Fig. 10 the
cyclotron masses of the SL-s and, for comparison, data from pure InAs
are plotted with the theoretical solid line. The agreement is quite
satisfactory except for two samples which have the narrowest periods.
In these, possibly the polarization of the GaSb layers contributes an
additional mass enhancement. Very recent measurements [31] on undoped
InAs-GaSb SL-s of various layer thicknesses give the first indications
of the semiconductor to semimetal transition. The carrier concentration
was found to increase from the 10^{16} to the 10^{17} range as the layer

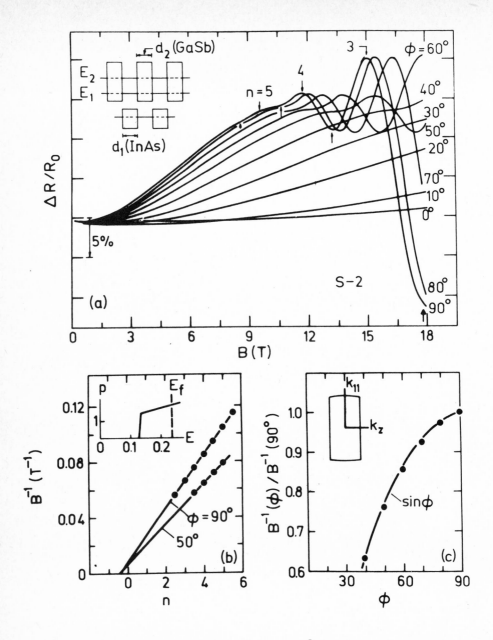

Fig. 9. a) InAs–GaSb SL magneto-oscillation at 4.2 °K. b) The plot of extrema fields vs. Landau index for two different angles. c) The shift in extrema as a function of angle between the field and the plane of the SL.

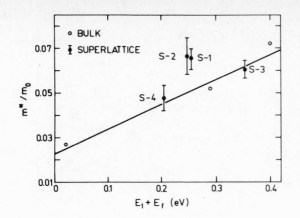

Fig. 10. Cyclotron effective mass vs. total energy,
the sum of ground subband energy E_1 and
Fermi energy E_f. The bulk InAs results
were taken from literature. Solid line
represents the expected dependence.

thickness was increased from below to above 100 Å. The thick layer samples show two sets of oscillations in the parallel conduction under the application of magnetic field, believed to be associated with electrons and holes, respectively. The effective masses, reduced from the temperature dependence of the oscillatory amplitudes, are ~ 0.04 and ~ 0.2 m_o for the two type of carriers.

Optical absorption measurements were performed [30] on SL-s of periods in the range of 40 to 70 Å. Our interest lay in investigating the dependence of the fundamental gap of Type II. SL-s on the period, as well as alloy composition, since the measured absorption edges provide a test for the subband theory in Type II. SL-s. The absorption curves for three samples with different alloy compositions and layer thicknesses are shown in Fig. 11. The samples represent different band-edge lineup configurations: large negative E_s, $E_s \sim 0$, and large positive E_s with no interaction between the VB and CB of neighbouring layers. The arrows show the gaps of the hosts and the calculated positions of the SL gaps, indicating excellent agreement with experiment. The absorption is weak and the SL gaps are narrower than that of their respective hosts. The latter is a striking feature of the Type II. band--edge lineup SL-s.

Next, to obtain E_s for pure InAs-GaSb, the absorption of several samples made of equal layer thicknesses was measured. In Fig. 12 the energy positions of the absorption edges are shown as a function of the layer thicknesses. The solid and broken lines are the calculated SL gaps for various values of E_s. Although some uncertainty exists, it is clear that the overlap is negative and falls in the vicinity of −0.15 eV. In a recent three-dimensional band calculation [32] of the InAs-GaSb SL agreement with our data was obtained by assuming again a negative E_s. The difference between the one-dimensional and, in this narrow range more reliable, three-dimensional treatment is insignifi-

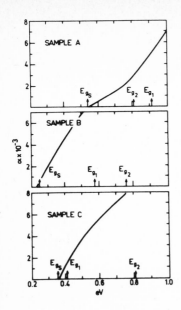

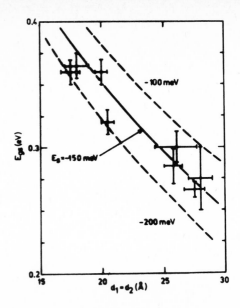

Fig. 11. Absorption coefficient vs. photon energy for three samples. Calculated energy gaps of the SL-s and the gaps of the hosts are marked with arrows.

Fig. 12. A comparison of experimental and theoretical energy gaps of InAs-GaSb SL-s as a function of the layer thickness. The energy overlap is indicated is used for parameter in the calculations.

cant in comparison to the expected [26] shifts in band-edge lineup if charge rearrangements had occurred. To our knowledge this is the first time that a negative overlap of the gaps has been determined at the interface of two semiconductors. The fact that the overlap can be negative has bearing in general on the theory of heterojunction interfaces.

ACKNOWLEDGEMENTS - Our work discussed here representes a group effort headed by L. Esaki. Partial support came from the Army Research Office, Durham, N.C.

REFERENCES

1. L.Esaki and R.Tsu, IBM J. Res. Develop. 14, 61 (1970)
2. P.J.Price, IBM J. Res. Develop. 17, 39 (1973)
3. R.Tsu and L.Esaki, Appl. Phys. Lett. 19, 246 (1971)
4. L.V.Keldysh, Fiz. Tverd. Tela 4 2665 (Sov. Phys.-Solid State 4), 1658 (1962)
5. R.A.Suris and V.A.Fedirko, Fiz. Tekh. Poluprovodn. 12, 1060 (Sov. Phys. Semicond. 12, 629) (1978)
6. G.H.Dohler, Phys. Status Solidi 52, 533 (1972)
7. J.W.Matthews and A.E.Blakeslee, J. Vac. Sci. Technol. 14, 989 (1977)
8. N.Holonyak, Jr., R.M.Kolbas, W.D.Laidig, B.A.Vojak, R.D.Dupuis and P.D.Dapkus, Appl. Phys. Lett. 33, 737 (1978)
9. G. A.Sai-Halasz, R.Tsu and L.Esaki, Appl. Phys. Lett. 30, 651 (1977)
10. L.Esaki, L.L.Chang, W.E.Howard and L.V.Rideout, Proc. 11th Int. Conf. on the Phys. of Semicond., Warsaw, Poland (1972)
11. L.Esaki and L.L.Chang, Phys. Rev. Lett. 33, 495 (1974)
12. R.Dingle, W.Wiegmann and C.H.Henry, Phys.Rev.Lett. 33, 827 (1974)
13. R.Dingle, A.C.Gossard and W.Wiegmann, Phys.Rev.Lett. 34, 1327 (1975)
14. R.Tus, L.L.Chang, G.A.Sai-Halasz and L.Esaki, Phys. Rev. Lett. 34, 1327 (1975)
15. P.Manuel, G.A.Sai-Halasz, L.L.Chang, C-A. Chang and L.Esaki, Phys. Rev. Lett. 37, 1701 (1976)
16. L.L.Chang, H.Sakaki, C-A.Chang and L.Esaki, Phys. Rev. Lett. 38, 1489 (1977)
17. G.A.Sai-Halasz, A.Pinczuk, P.Y.Yu and L.Esaki, Surf.Sci. 73, 232 (1978)
18. G.A.Sai-Halasz, A.Pinczuk, P.Y.Yu and L.Esaki, Solid State Commun. 25, 381 (1978)
19. H.Sakaki, L.L.Chang and L.Esaki, Physics of Semiconductors 1978, Conference Series 43 The Institute of Physics Bristol and London, Edited by B.L.H.Wilson (The Institute of Physics, Belgrave Square, London) (1979)
20. R.Dingle, Surf. Sci. 73, 229 (1978)
21. A.C.Gossard, P.M.Petroff, W.Wiegmann, R.Dingle and A.Savage, Appl. Phys. Lett. 29, 323 (1976)
22. R.Dingle, H.L.Stormer, A.C.Gossard and W.Wiegmann, Appl. Phys. Lett. 33, 665 (1978)
23. G.W.Gobeli and F.G.Allen, in Semiconductors and Semimetals, ed. by R.K.Willardson and A.C.Beer (Academic, New York) Vol.2, 263 (1966)
24. W.A.Harrison, J.Vac.Technol. 14, 1016 (1977)
25. H.Sakaki, L.L.Chang, R.Ludeke, C-A.Chang, G.A.Sai-Halasz and L.Esaki, Appl. Phys. Lett. 31, 211 (1977)
26. W.R.Frensley and H.Kroemer, Phys. Rev. B16, 2642 (1977)
27. G.A.Sai-Halasz, L.Esaki and W.A.Harrison, Phys.Rev. B18, 2812 (1978)
28. C-A.Chang, R.Ludeke, L.L.Chang and L.Esaki, Appl.Phys.Lett. 31, 759 (1977)
29. H.Sakaki, L.L.Chang, G.A.Sai-Halasz, C-A.Chang and L.Esaki, Solid State Commun. 26, 589 (1978)
30. G.A.Sai-Halasz, L.L.Chang, J-M.Welter, C-A.Chang, L.Esaki, Solid Sate Commun. 27, 935 (1978)
31. L.L.Chang, G.A.Sai-Halasz, N.J.Kawai and L.Esaki, 6th Annual Conf. on the Phys. of Comp. Semicond. Interfaces, J.Vac.Sci.Technol., to be published (1979)
32. A.Madhukar, N.V.Dandekar and R.N.Nucho, 6th Annual Conf. on the Phys. of Comp. Semicond. Interfaces, J.Vac.Sci.Technol., to be published (1979)

THE LOCALIZED STATES OF INTERFACES AND THEIR PHYSICAL MODELS

J. Giber

Physical Institute of the Technical University Budapest

H-1502 Budapest, P.O.Box 112, Hungary

INTRODUCTION

The better understanding of the electronic structure of real surfaces has a scientific and practical importance as well. From theoretical point of view this question involves the problem of every system
where the long range order - being the basis of the early development
of solid state physics - is broken or lacking.

The development of useful theoretical models and methods requires
a large amount of reproducible experimental knowledge. The Si/SiO_2
system is one of the most extensively investigated systems because of
its importance in MOS devices. During the past decade many empirical
data have been accumulated, and well controlled interfaces are available. However, little is known about the origin and the physical
nature of the localized electronic states of the system.

In the first part of this paper after an empirical classification
of localized electronic states and a review of experimental results on
properties of them in the Si/SiO_2 system, we shall suggest a new
classification of localized states based on the knowledge about the
real structure of the oxide (considering the interface and the oxide
together as a disordered system).

In the second part we shall show, after a review of calculational
methods applicable for amorphous materials and for point defects in
crystals, that the qualitative picture for the electronic structure of
such a system, taking the results of preliminary calculation into
account, agrees well with the experimental facts.

I.

1. EMPIRICAL CLASSIFICATION AND PROPERTIES OF LOCALIZED STATES

In a real MOS structure there are localized electronic states in the Si bulk, in the SiO_2 bulk and in the interfacial transition region. Since the effect of all of these is present in the experiments, the problem of the interface extends to all such localized states of the system.

The classification of the localized states was carried out using their characteristics shown in experiments. In early papers many workers had their preferred but different terminologies for the same type of states.

An empirical classification of localized electronic states in a non-ideal MOS structure was given in the recent, very comprehensive review of Cheng [1] as follows (Fig. 1).

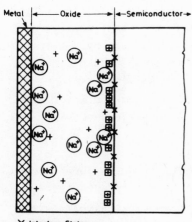

X Interface States

⊞ Fixed Oxide Charges

(Na⁺) Mobile Ions

+ Ionized Traps   Fig. 1.

Y.C.Cheng Progr. Surf. Sci, 8, 181 (1977)

Fig. 1.

1. Interface states
   - localized at or very near to the interface such that charge transfer between the interface and the Si bulk can take place readily;
   - strong dependence on the surface potential;
   - amphoteric nature.

   Other terminologies in the literature: fast interface states, fast surface states, fast states, surface traps, surface states...

2. Slow trapping states
   - similar to interface states;
   - generated at Si-SiO_2 boundary by the prolonged influence of a large electric field at sufficiently high temperature.

3. Ionized traps
   - distributed throughout the SiO_2;
   - generally positive charges
   connection with: damages caused by radiation or contamination by certain impurities.

4. Fixed oxide charge
   - positive charges in the oxide;
   - intrinsic to the Si/SiO_2 system;
   - located within 200 Å to the interface;
   - inable to exchange charges with the Si;
   - stable in a large electric field and at elevated temperature.

5. Mobile ions
   - due to contamination, extrinsic impurities;
   - can be drifted across the oxide by applying an electric field
     even at room temperature.

The empirical way of distinction among the states in this classification can be visualized on the well known high-frequency capacitance-
-voltage (CV) measurement (Fig. 2). Varying the gate voltage from in-
version to accumulation the change in the position of the Fermi level at the interface causes a change in the occupa-
tion of trapping localized states, followed by a non-parallel change in the AB section (AB $\longrightarrow$ A'B'). The distance of the traps from the interface can be determined varying the frequency, applying which the trap can be activated. This is

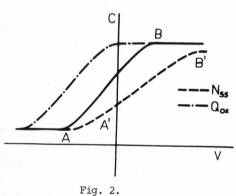

Fig. 2.

the basis of the differentiation among 1., 2., 3. Furthermore there
are permanently charged centres (4., 5.), which are not affected by the
change of the voltage. The potential field of these is simply added to
the gate voltage causing a parallel shift of the ideal C-V curve. A
further distinction between 4. and 5 is possible corresponding to the
behaviour in bias-temperature stress tests.

The presence of the mobile ions in the system can be well controlled, and their origin is obvious while the experimental knowledge
about 2. is not enough for extended discussion. Therefore in this
article we concentrate on classes 1., 3. and 4. The term "interface
state" will be used in a broad sense, i.e. for any electronic state,
the wave function of which has a maximum amplitude at, or near the interface,
and vanishes at sufficiently large distance from the interface. The
origin of fixed oxide charge and that of ionized traps is not clear.
The experimental distinction is not unambiguous; it is possible that
they have common characteristics. Therefore we will treat these

together as *oxide traps*. They will be regarded as centres in the bulk oxide providing localized states with energy appropriate for communication with the silicon surface charge distribution by a certain mechanism (not via charge transport by all means).

EXPERIMENTAL RESULTS ON THE PROPERTIES OF THE INTERFACE STATES

1. Oxidation conditions
- best (most clean and ideal) interface can be produced by thermal oxidation in a mixture of O_2 and Cl_2;
- the higher the water content of the oxidi÷ing ambient, the lower is the density of interface states (N_{ss}).

2. Annealing
The effect of heat treatment in different ambients can be seen in Fig. 3. We note, that the effect of He (and similarly N_2) is attributed to the associated water content (in the descending branch) while the increasing effect is similar to that of vacuum.

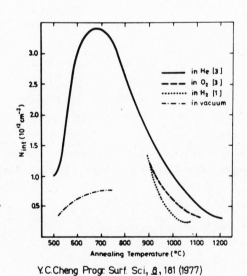

Y.C.Cheng Progr. Surf. Sci., **8**, 181 (1977)

Fig. 3.

3. Orientation dependence
- The density of interface states is at about $11.8 \cdot 10^{14}$ cm^2 for (111); $9.6 \cdot 10^{14}$ cm^2 for (110) and $6.8 \cdot 10^{14}$ cm^2 for (100) interfaces [4].

4. Irradiation and ion implantation
- Both processes result in radiation damage (band breaking and distortion) which is followed by an increase in the number of interface states [5].
- The additional interface states due to dopant ions implanted in the vicinity of the interface have almost the same energy as in the bulk [4].

5. Energy distribution of interface states in the gap

No general method is available for measuring the energy distribution throughout the whole energy gap, the results of two or more methods must be combined. The measured distribution profiles show a broad minimum in the midgap and the concentration increases toward the band edges. The works [6,7,8], using conductance or quasistatic techniques, reported a continuous increase at the vicinity of the band edges. Other methods proposed by Gray and Brown [9] give peaks close to the band edges [9,93]. However the spacings of these are not reliably determined. If there are peaks, they must be closer than 0.1 eV to the band edges. Further investigation is needed in this field.

6. Charge state of interface states

Macroscopic measurements show that the interface states are not charged when the levels are not occupied. This can be interpreted in terms of neutral centres [9] or in terms of equal number of negatively and positively charged centers [10]. While the former corresponds to acceptor states near the conduction band and donor states near the valence band, the latter assumes the contrary of this. The decision would be possible with capture cross section measurements. However, these are also not unambiguous because the shape of energy distribution must be assumed. If an energy distribution increasing monotonously toward the band edges is accepted, such a measurement supports the neutral center model [11]. On the other hand, other measurements [12] making use the statistical model of [13] yield donor states in the vicinity of the conduction band, and a much lower density of acceptor states near the valence band.

EXPERIMENTAL KNOWLEDGE ABOUT OXIDE TRAPS

The oxide traps are located throughout the whole oxide with a density of about 10^{10}-10^{13} cm^{-2}, but unfortunately the distribution in space is not known.

1. Oxidation condition
- The "dry" oxidation provides less oxide charge (Q_{ox}) than the wet oxidation, but the effect of water content is small [14].
- Higher oxidation temperature results in a lower Q_{ox} [14].

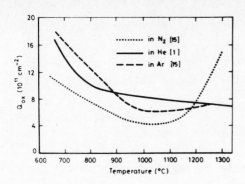

Fig. 4.

2. Annealing

We note the deviation from annealing effects on interfaces states.

3. Energy and capture cross section

- Hole trapping: Table I.
- Electron trapping: Table II.

Table I.

Energy (eV above the SiO$_2$ valence band)	σ (cm^2)	Density of states (cm^{-2})	Assignment	Ref.
	$1.7 \cdot 10^{-14}$	$1.7 \cdot 10^{12}$	neutral or compensated centers	[16]
	$1.0 \cdot 10^{-13}$	$7.0 \cdot 10^{11}$		
	$1.9 \cdot 10^{-14}$	$2.0 \cdot 10^{12}$		[16]
	$3.1 \cdot 10^{-13}$	$1.8 \cdot 10^{12}$	compensated negatively charged centers	[17]
4-5			oxygen vacancies	[18]

Table II.

Energy (eV below the SiO$_2$ conduction band)	σ (cm^2)	Density of states (cm^{-2})	Assignment	Ref.
	$2 \cdot 10^{-15} - 5 \cdot 10^{-20}$		Na$^+$	[19]
	$1 \cdot 10^{-17} - 2 \cdot 10^{-28}$	$2 \cdot 10^{12}$	water content	[19]
2.5		10^{10}		[20]
2	$1.3 \cdot 10^{-12}$	$3 \cdot 10^{14}$	positively charged Coulomb center	[21]
2.4	10^{-14}	$1 \cdot 10^9$	Coulomb attractive	[22]
2.1			water content	[23]

2. STRUCTURE OF THE Si/SiO$_2$ SYSTEM

All the physical models of these states have set out from the real structure, i.e. from the arrangement of the atoms at and near the interface. Therefore let us turn to the structural informations on the Si/SiO$_2$ system. The thermal oxidation proceeds in three steps [24] as follows:

1) The transfer of oxidizing species into the oxide already formed.
2) The diffusion of the species through the oxide.
3) The reaction of oxygen with silicon in the interface region.

Due to the oxygen "interstitials" in the silicon lattice at the interface, the Si-Si bonds break when the concentration of oxygen increases and only a part of these are bridged by the oxygen atoms present. Therefore a transition layer will be found with excess number of silicon atoms. On the other hand, the diffusing oxygen interstitials in the SiO$_2$ have an increasing distribution toward the free surface. In this way a stoichiometric polarization will be formed. This kinetic model is supported by RBS [25/a] and SIMS [26] measurements.

The crystallographic structure of the oxide layer formed this way is that of vitreous silica (e.g. [27,28]). This is determined by the short range order of SiO$_4$ tetrahedra. No crystallites were found in the oxide layer, and the interface was shown to be defectless [27].

The structure of the interface i.e. the stoichiometry, smoothness and the width of the SiO$_x$ transition region was the subject of extensive research in recent years (Table III).

The conclusion which can be drawn is the existence of a transition layer with a thickness of about 10 Å containing a number of $\sim 10^{15}$ excess silicon atoms, and the shape of this region is straight in ~ 20 Å.

In addition to the intrinsic structure of the Si/SiO$_2$ system we note the "almost intrinsic" hydrogen contamination, which is accumulated at the interface [38,39].

3. PHYSICAL MODELS OF INTERFACE STATES AND OXIDE TRAPS

Now in possession of the required experimental knowledge, let us review the models proposed for interface states and oxide traps. As since the appearance of Cheng's review new developments did not arise, we recall the summary of models presented there [1] (Table IV-V).

Table III.

Method	Stoichiometry (number of excess Si atom per cm^2)	Smoothness (width of the interface) (Å)	Transition layer (width of SiO$_x$ layer) (Å)	Ref.
RBS high energy	6·10^{15}			[25a]
low energy	14·10^{14}		15-20	[25b]
Wetting of SiO$_2$ with water			30	[29]
ESCA			20	[30]
ESCA		15-20		[31]
XPS	10^{15}		14	[32]
Auger		20-30		[33]
Auger		20	6	[34]
Photoemission			4	[35]
High resolution el.microscopic		4	10	[36]
Electron mobility			5	[37]
SIMS			50	[93]

Among the models of interface states (Table IV) the oxygen model
of Cheng and the three layer model of Litovchenko should be o-mitted
since they contradict experimental facts about the structure of the
system. In 1968 Goetzberger [52] proposed a model in which the interface states are caused by fixed charge in the oxide. As the experiments
of Krafcsik [53a,b] showed, there is a connection between the fixed
charge and the surface states. A more rigorous calculation led to the
conclusion that the simple interpretation proposed by Goetzberger
could not be valid (see Table IV) [65,52].

The Si vacancy - interface stress model of Lane (Table V) seems
to belong rather to interface states than oxide traps though this
author compared his experiences with experiments upon oxide charge.
However, the model is not supported by other measurements.

Table IV.

Model	Origin	Major success in explanation of	Defficiency
Effect.mass appr. or Goetzberger model [40]	Coulomb like 1s states. Caused by fixed charges in the oxide near the interface	Peaks position is in accordance with some experiments	Qualitative only. The role of the boundary conditions is not clear; 1s like state is not possible; the image charge is not regarded.
Interface disorder Révész [41]	Associated with disorder in the interface region due to thermal oxidation	Dependence from oxidation conditions and orientation	Qualitative only
Misfit dislocation model Neumark [42]	Associated with the misfit dislocation series of the interface	Peak in the mobility--vs-gate voltage plot; orientation dependence; measurable interface state band	Amphoteric nature is not explained; applicable in special samples only (dislocations can be emitted rather well)
Oxygen model Cheng [43]	Due to groups of oxygen atoms weakly bonded on Si surface	Wide variety of experimental facts (qualitativity); especial amphoteric nature	No such oxygen groups are present at the interface; after preliminary calculation the states associated with oxygen dangling bonds are shifted into the bonds of Si [46]
Three layer model Litovchenko [44]	defective Si | 2 dim. SiO_2 cryst. | amorphous SiO_2	Wide variety of properties	No such crystalline layer at the interface exist
B_2 centre Hickmott [45]	Due to B_2 centres in SiO_2 near the interface	Annealing properties and a part of the energy distribution	Origin is not known; a partial possibility only
Vacancy model Iizuka,Sugano [46]	Oxygen vacancy on the interface causes localized levels in the gap of Si	Orientation dependence; amphoteric nature; oxidation conditions and annealing properties	Semiquantitative and not enough elaborated
Excess Si model Deal [24]	Associated with dangling bonds of excess Si atoms in the transition region	Dependence on oxidation condition annealing properties	Qualitative only

Table V.

Model	Origin	Major sucess in explanation of	Main feature
Trivalent Si model Révész [47]	Breaking of oxygen bridges due to hydrogen contamination	Donor levels; annealing properties	Qualitative only
Excess Si model Deal [14]	Breaking of Si bonds due to oxidation process	Annealing properties	Qualitative only
Interfacial stress Lane [48]	Vacancies in Si in interaction with interface stress	Correlation with oxidation temperatures	Contradicts some experimental facts (note, that it is not clear whether this is a model of surface states or oxide traps)
Oxygen vacancy model Thomas, Young [49]	Due to oxygen vacancy in SiO_2	Positive charge by an appropriate gap level [77]; almost all annealing properties and the dependence on oxidation process	Semiquantitative
E_1' centre model Fowkes [50]	E_1' centre in SiO_2	Annealing properties	No origin
Dipole layer Kwon [51]	Due to asymmetric el. dipole at the interface	Flat band shift	Annealing properties are not explained

All the remaining models are based on defect centres (misfit dislocations; B_2 centre; vacancy at the interface; dangling Si bonds at or near the interface; oxygen vacancy in SiO_2; E_1' centre) or on band-direction distribution (disorder, dipole). Note, that while the origin of B_2 centre is not known, the E_1' centre was identified as an asymmetrically relaxed oxygen vacancy in SiO_2 [54]. These, and irradiation experiments support the validity of oxygen vacancy model.

4. CLASSIFICATION OF THE LOCALIZED STATES ACCORDING TO THE STRUCTURE
   OF THE Si/SiO$_2$ SYSTEM

All the models mentioned, pick out a particular characteristic of
the structure and try to explain a part of the properties. Most of
them are only speculative in nature and any quantitative justification
regarding the energy levels produced by the model is lacking. An other
problem is that the indefiniteness of the samples must be kept in mind.
An amorphous structure prepared by diffusion among not well controlled
conditions can have a great number of various extrinsic and intrinsic
defect centres. Since the various experiments were not carried out on
the same sample, the results can not be identical. Consequently, the
models based on individual centres, do not have to explaine all experimental facts, and on the other hand their effect do not have to be
present in all samples.

Some of these problems originate from failures of the early MOS
technology. As today the creation of very smooth interface with thin
transition layer and low extrinsic defect concentration as well as the
preparation of a clean and little intrinsic defect-containing oxide is
possible, we can turn to the investigation of an "intrinsic" Si/SiO$_2$
system, postpone the problem of extrinsic impurity contamination, of
dislocations etc.

An overall satisfactory picture of the problem can be obtained,
considering the whole of such a structure. To face the issue is rather
disencouraging. We have to calculate the electronic structure of an
amorphous system containing point defects and joining with more or
less mismatch to a system with ordered crystal structure. This requires
three fields of solid state physics, which for the time being are under
development. But, before introducing the methods going to be applicable
perhaps in the future only, let us draw some qualitative conclusion
about the nature of localized states in such a system.

The early concept of interface states due to disturbation of
periodicity of silicon on the interface can be regarded as antiquated.
The band picture of solids is consistent with a short range order too
(see later), and the abruption of the system causes discrete gap levels
associated with dangling bonds. In the case of an interface with a
flexibile amorphous material as SiO$_2$, the dangling bonds are mostly
saturated. This new bonds differ from those of the bulk, but this
difference causes localized states lying in the Si bands only. This
statement was justified by simple calculations [55,46], but can be
seen simply from the relation of the band structure of Si and SiO$_2$.
Therefore it seems to be more convenient to regard the problem from

the side of the oxide, i.e. the structure under consideration is a
disordered system, in main part of which the offdiagonal disorder is
dominant (i.e. the different linkages of the SiO_4 units) while a
diagonal disorder appears in the interface region (i.e. linkage of a
SiO_4 unit with whether a SiO_4 or Si_4 unit). In the meantime point defects can occur in the whole system.

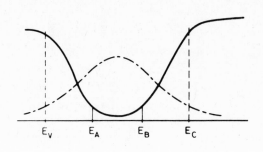

Fig. 5.

The electronic structure
of short range ordered systems is the subject of extensive development in recent
years with particular interest
to amorphous semiconductors.
A qualitative picture can be
given about glassy materials
after Mott [56]. The energy
diagram of the electronic
states in a short range or-
dered system is shown in Fig. 5. A continuum of extended states exists below and above the mobility edges corresponding to valence band
edge and conduction band edge respectively. The tailing of both bands
can be observed till E_A and E_B resp. corresponding to localized states
originating from *bond bending* and distributed uniformly in the whole
structure. (Conduction can take place among these states by hopping.)
The $E = E_A - E_B$ separation corresponds to a pseudogap with low density of states. In addition several peaks can occur in the range of
this gap due to point defects intrinsic or extrinsic to the structure.

For the case of silica, the tails were found to be narrower than
kT [57]. This indicates the correctness of taking the electronic structure of SiO_2 from band structure calculations of α-quartz or β-crys-
toballyte. This is supported by many other experimental data and theoretical considerations [58,59,60,61,62] too.

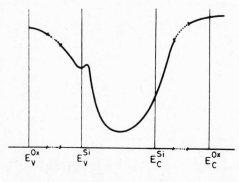

Fig. 6.

In the transition region
a similar picture can be expected. Though detailed description
can only be expected from a calculation, we believe that extended states over the entire
system are present, accompanied
by band tails stretching
deeper in the gap as in the case
of pure oxide. The broken bonds
and vacancies as well as extrin-

sic impurities cause additional amounts of states at certain energies. The speculative picture of such a state distribution agrees well with the observed ones (Fig. 6). This approach suggests a new natural way of classification (Table VI).

Table VI.

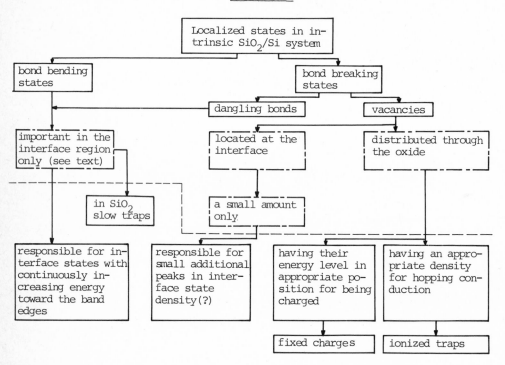

The distribution of vacancies can increase toward the interface because of the interface stress. This can explain the known location of oxide charges [63].

The correspondance between the parts above and below the dashed line is of course somehow arbitrary, but taking the nature of the Si/SiO$_2$ structure into account, a qualitative agreement can be easily shown. By all means all the previous models based on intrinsic features of the system can be unambiguously set in this classification. However, we regard these in a general picture and show a way for carrying out a quantitative analysis, which thus far was lacking. Of course, only such a calculation can give the possibility to make a detailed comparison of the models with the experimental data accumulated previously. Let us now collect the computational methods available. Most of them of course were not applied so far to the concrete field of interest. However, we wish to take the opportunity to classify these methods worked out in a large scale of various fields of solid state physics, to summarize their essence and to show the way of application in our problem.

II.

1. METHODS OF CALCULATION OF THE ELECTRONIC STRUCTURE OF BOND BREAKING STATES

As mentioned above, the electronic structure of the oxide only slightly differs from that of crystalline forms of SiO_2. This fact can be used to simplify the calculations of energy levels of point defects locating in the oxide bulk. Therefore the techniques of calculating deep levels in crystals can be applied. These methods were summarized by Pantelides [64]. Somewhat supplementing his review, the practically applicable methods in the case of a vacancy are the followings:

A) Green function-methods

Let be

$$H\psi \equiv \left[-\frac{\hbar}{2m} \Delta + \sum_{\alpha} V(\bar{r}-\bar{R}_{\alpha}) \right] \psi = -\frac{\hbar}{i} \frac{\partial \psi}{\partial t} \tag{1}$$

the Schrödinger equation of the perfect crystal. Than the Green-function $\tilde{G}$ of the system is defined by the solution of the

$$\left[\frac{\hbar}{i} \frac{\partial}{\partial t} - \frac{\hbar^2}{2m} \Delta + \sum_{\alpha} V(\bar{r}-\bar{R}_{\alpha}) \right] \tilde{G}(\bar{r}, \bar{r}'; t, t') = \delta(\bar{r}-\bar{r}')\delta(t-t') \tag{2}$$

inhomogeneous equation. The Fourier transform of $\tilde{G}$ in terms of energy is

$$\tilde{G}_{\pm}(\bar{r}, \bar{r}'; E) = \sum_{n} \frac{\psi_n^*(\bar{r}') \psi_n(\bar{r})}{E-E_n \pm i\varepsilon} , \tag{3}$$

where E_n-s are eigenvalues of the stationary Schrödinger equation and $i\varepsilon$ in the denominator is needed to define the sense, in which the pole at $E=E_n$ has to be taken. The $\pm$ signs correspond, to outgoing and ingoing waves respectively. (3) can be written as

$$\tilde{G}_{\pm}(\bar{r}, \bar{r}'; E) = \sum \psi_n^*(\bar{r}')(E-H\pm i\varepsilon)^{-1} \psi_n(\bar{r}) \tag{4}$$

where the operator in the right hand side is the resolvent operator of (1) and hereafter referred as Green's operator G. Certain authors inaccurately call that as Green-function, too. In the energy gap, where H does not posess any eigenvalue, G unambiguously equals to $(E-H)^{-1}$

One can compute the density of states $n(E)$ with the help of the Green function [65],

$$n(E) = \frac{i}{2\pi} \int \left[\tilde{G}_+(\bar{r}, \bar{r}'; E) - \tilde{G}_-(\bar{r}, \bar{r}'; E) \right] d^3 r' \tag{5}$$

or by the Green operator [64]

$$n(E) = -\frac{1}{\pi} \, \text{Im} \, \text{Tr} \, G. \tag{6}$$

Let be $\{\varphi_n\}$ a complete set of functions, then (6) can be written as

$$n(E) = \frac{1}{\pi} \, \text{Im} \sum_\mu G_{\mu\mu}(E) \tag{7}$$

where $G_{\mu\mu}(E)$ is the matrix of G in the $\{\varphi_n\}$ representation.

Now, the methods can be devided into two groups: calculation of the Green function of the perfect crystal and taking the vacancy potential as a perturbation, or the Green function of the defective crystal directly. The former ca be called as perturbation methods, while the latter as cluster methods both in a generalized sense. (The calculation of the defective Green function can be extended over a finite number of sites only.)

A.1 Perturbative methods

These methods are based essentially on the Koster-Slater technique [66]. In the formulation of [64] it can be introduced as follows. Let be

$$H_\psi \equiv (H_o - V)\psi = E\psi \tag{8}$$

the Schrödinger equation of the imperfect solids, V being strongly localized perturbation. Let G^o the Green's operator of the perfect crystal. Than for states *in the gap* $G^o = (E-H)^{-1}$ and therefore

$$(E - H_o)\psi = V\psi \tag{9}$$

$$\psi = G^o V \psi \tag{10}$$

$$(1 - G^o V)\psi = 0 \, , \tag{11}$$

which has a nontrivial solution only when

$$\det(1 - G^o V) = 0 \quad . \tag{12}$$

Within the band E can be degenerated with one of the energies E_n of the perfect crystal, therefore the general solution of (9) is

$$\psi_n = \psi_{nk}^o + G^o(E_n) V \psi_n \tag{13}$$

and here

$$G^o(E_n) = \lim_{\substack{\varepsilon \to 0 \\ E \to E_n}} (E - H_o + i\varepsilon)^{-1} \tag{14}$$

Let us define the Green's operator of the imperfect crystal as

$$G(E) = (E - H)^{-1} \tag{15}$$

(E is a complex variable, and its small imaginary part will be let go to zero in an appropriate way.) By using the identity

$$G(E) = -\frac{d}{dE} \ln G(E) \tag{16}$$

(6) can be rewritten as

$$n(E) = \frac{1}{\pi} \operatorname{Im} \frac{d}{dE} \ln \det G(E) \quad, \tag{17}$$

similarly

$$n^o(E) = \frac{1}{\pi} \operatorname{Im} \frac{d}{dE} \ln \det G^o(E) \quad. \tag{18}$$

Now using the definition of G^o in (15)

$$G = \frac{1}{E - H_o - V} = \frac{1}{E - H_o} + \frac{\frac{1}{E - H_o}}{E - H_o - V} = G^o + G^o V G \tag{19}$$

we get Dyson's equation, which can be solved by

$$G = (1 - G^o V)^{-1} G^o \quad. \tag{20}$$

Using this in (17) we get

$$n(E) = n^o(E) + \Delta n(E) \tag{21}$$

where

$$\Delta n(E) = -\frac{1}{\pi} I_m \frac{d}{dE} \ln \det(1-G°V) \tag{22}$$

is the change in the density of states, and can be calculated directly.

No let us return to the gap. Expanding $|\psi\rangle$ in terms of any complete set of functions to be $|\psi\rangle = \sum_n F_n |\varphi_n\rangle$ and using Eq.(11)

$$\sum_{n,n'} \langle \varphi_{n'} | (1-G°V) | \varphi_n \rangle F_n = 0 \quad , \tag{23}$$

which has nontrivial solution if the determinant of the matrix vanishes.

The use of different basis sets is possible. Koster and Slater proposed Wannier representation or the set of orthonormal Bloch waves constructed from atomic orbitals. Since the work with these requires too much computational effort, Bernholc and Pantelides [67] and Pantelides [68] suggested the use of a simple LCAO (hybrid) basis, making use of the localized nature of V. They calculated the number of matrix elements of the operators $(1-G°V)$ according to the range of V. In [68] V was determined as a self consistent pseudo-potential. Essentially the same way was followed by [69] and [70], except of using

$$\det(1-V_2 G°V_1) = 0 \tag{24}$$

when $V_2 V_1 = V$ and of using a complete set of localized functions as a basis set.

All these methods were applied to deep levels in diamond and zincblende type semiconductors with succes.

A.2 Cluster methods

These methods go act from the imperfect crystal's Green operator. The equation (7) of density of states is rewritten as

$$n(E) = -\frac{1}{\pi} I_m \sum_\mu G_{\mu\mu}(E) = \sum_\mu n_\mu(E) \quad , \tag{25}$$

where

$$n_\mu(E) = -\frac{1}{\pi} I_m G_{\mu\mu}(E) \tag{26}$$

is a local density of states. The calculations are proceeded for an appropriate number of sites to achieve convergent results for the density of states of the perfect crystal, than by removal of the central atom, $n_\mu(E)$ is recalculated in the vicinity of the defect.

Kauffer et al. [71] calculated $G_{\mu\mu}(E)$ by the use of the continued fraction method. This is a recursion process

$$G_{\mu\mu}(E) = E - a_1 - \cfrac{1}{E - a_2 - \cfrac{b_1}{E - a_3 - \cdots}}$$

$\qquad(27)$

where the parameters a_i and b_i are determined by complicated formulas from a tight binding Hamiltonian.

An another solution was made by Joannopoulos and Mele [72]. They introduced an averaged Green-function satisfying a set of coupled linear differential equations in energy, which can easily be solved for even very large systems.

B) Secular matrix methods

In contrary to Green function methods these methods solve the secular equation directly. There are two classes of these. Methods from the first group retain a certain part of the translational symmetry, i.e. they carry out a band structure calculation for a larger unit cell. These are the periodic cluster or supercell methods. The other kind of cluster methods utilize large truncated clusters. (This was the original form of the cluster models.) While in this latter the main problem is the defect-surface interaction and therefore it is suitable to qualitative description only, in the former this is substituted by defect-defect interactions. This has obviously much smaller effect and therefore with application of self consistent methods one is able to get quantitive predictions.

The avoiding of underisable defect-defect interactions is possible by increasing the cluster size. However, the convergence with cluster size is a hard question. Direct proof is not possible due to the limited computer capacity. Previous estimates on the localization of vacancy wave function are available from not convergent, and mainly not self consistent model calculations only.

A cluster of 54 atoms was used for calculating a vacancy in silicon by means of self consistent pseudopotential band theory by Louie et al. [73]. They found that an unrelaxed ideal vacancy changes the charge distribution in its direct neighbourhood only.

However, a band structure calculation especially for such a large unit cell is rather time consuming. Therefore Evarestov et al. [74] suggested the calculation of the $k = 0$ state only with neglecting the interaction of this with all other k states. (Note that for a perfect large unit cell, k = corresponds to a selected number of high symmetry points of the primitive unit cell Brillounin-zone: Γ, X, L, ... etc.). This procedure, however sets up the problem of convergence of the perfect cluster energies to the band structure.   Taking of Chadi and Cohen's representative k-point theorem [75] into consideration is encouraging in this respect. The application of self consistent semi-empirical LCAO-MO techniques solving the problem is subject of the next paper [76].

From all the methods mentioned only two were applied to SiO_2. A truncated cluster approach was made by Yip and Fowler [54], and a primitive form of supercell method (using the primitive cell and EHT Hamiltonian) by Bennett and Roth [77]. However, important semiquantitative conclusions were drawn. The oxygen vacancy in neutral state has a doubly occupied level above the valence band and an empty one below the conduction band. Their separation corresponds to the transition of the E_1' centre. In positively charged state the lower state has a position above the midgap, and therefore remains permanently charged in the MOS system. Bennett and Roth carried out calculations on various bond breaking states and on vacancy. A further improvement of these calculations however seems to be difficult because of the large primitive cell of various forms of SiO_2. Therefore perhaps the use of approximate Hamiltonians may be necessary [60,61,62].

2. METHODS OF CALCULATION THE ELECTRONIC STRUCTURE OF BOND BENDING STATES

Now let us turn to the interface again. The theory of surfaces was developed in an intensive manner in recent years. However, the results of such calculations as those of Pandey and Phillips [78] and Appelbaum and Hamman [79,80] can be transformed to interfaces between crystalline materials with similar structure only, because they make use of the states of the bulk band structure. A more convenient approach is that of Louis [81], who kept short range order only. This method makes possible also the inclusion of point defect to the interface.

In this method the bulk of the solid is described in the Bethe (or tree) lattice approximation with the bulk Green's operator, which includes four hybridized sp^3 orbitals and all types of σ and π interactions [92] but only first-neighbours interaction is considered. The interface was studied by connecting one side of the crystal to the bulk and the other side to the other medium in the interface. They got equations for the transfer matrices at the interface, which involve the matrix elements of the Green's function.

However, a general picture for the Si/SiO$_2$ interface can only be expected, if we use the theories of disordered systems. These theories were elaborated for disordered alloys, while in the recent years some effort was carried out for application to amorphous semiconductors substituting the atoms on the sites with clusters representing the units of the short range order. Therefore we briefly review the origine of these methods.

Edwards and Beeby [82,83] developed the so-called *multiple scattering theory (MST)* of disordered systems. The motion of an electron in the field of fixed centres is regarded as a succession of elementary scatterings on the randomly distributed atomic scatterers, which scatterings are then averaged over all atomic configurations. The average density of states can be written as

$$\langle n(E) \rangle = \int \rho(\bar{k}, E) d^3 k \,, \tag{28}$$

where $\rho(\bar{k}, E)$ is the mean square probability of finding and electron having an energy E and momentum k. This is the Fourier transform of

$$\rho(\bar{r} - \bar{r}_i, E) = \frac{1}{2\pi} \langle \tilde{G}_+(\bar{r}_i, \bar{r}_i^{\,\prime}, E) - \tilde{G}_-(\bar{r}_i, \bar{r}_i^{\,\prime}, E) \rangle \tag{29}$$

The effect of disorder is that to continuously distort the $\rho(\bar{k}, E)$ surfaces. The magnetic moment and specific heat of binary alloys was calculated within this scheme.

Keller [84] employed the method to analyse the electronic density of states in crystalline and amorphous materials. He investigated what is the smallest cluster retaining the characteristics of the real amorphous material, and studied the effect of boundary conditions. The average density of states could be divided into three parts:

$$N(E) \equiv \langle n(E) \rangle = N_o(E) + N_1(E) + N_2(E) \tag{30}$$

where N_o is a free electron contribution, N_1 is the contribution of scattering in each cluster and N_2 is that of scattering between clusters. This last term corresponds to the fact, that a cluster is not isolated, i.e. N_2 is the boundary correction.

In the case of negative energies for clusters of medium and large size all the relevant information will be in terms of the single cluster (the free electron term also disappears). The short range order is then the only relevant information about the geometry of the material. This special case was not throughout discussed in the paper because this energy region was out of interest.

A second special case which is particularly significant for the
d electron bands in liquid and solid transition metals (*Keller and
Jones, 1971*) [85] is when the energy is near a resonance or antiresonance. In this case the long range order does not give any special
contributions. Finally the boundary correction will be less important
in the region of large electron density of states where strong scattering and probably resonances will make the major contribution, but
the boundary correction will be more significant in the energy gap
regions.

The studied boundary conditions were:
1. the cluster is in an average outside potential equal to the potential between scatterers in the cluster;
2. the cluster isolated from the crystal;
3. Born-Kármán periodic boundary conditions as in a superlattice of
   identical clusters.

Calculations were made for clusters of 1, 2, 6, 8, 10, 18, 30 C
and Si atoms characterised by muffin tin potential in a series of different configurations. In all cases the low density of states region
lies at higher energies than the forbidden gap found in the
perfect crystal. Fig. 7 shows the density of states as a function of
electron energy for the case of the first boundary condition (8 atom
cluster).

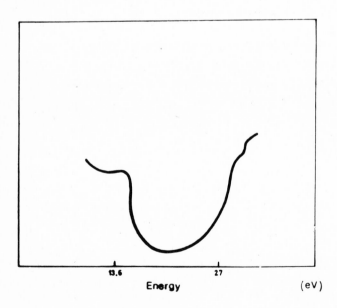

Fig. 7.

No evidence for states in the centre of the pseudo energy gap region was found in the case of "breaking" some of the bonds in the cluster. Different local configurations having a similar number of nearest neighbours and mass density gave essentially the same results for $N(E)$, but increasing the volume of a cluster with fixed number of atoms destroyed the energy gap. Introducing topological defects (in which the coordination number rule is not satisfied) by removing an atom no significant change in the energy gap region was produced, contrary to the case of geometrical defects (when any of the nearest neighbours distances or bond angles lies outside their specified range).

The qualitative results of MST can be improved by the coherent potential approximation (CPA). In CPA each scattering center is regarded as being embedded in an effective medium, characterized by a coherent potential $V_c(k,E)$, which is a complex quantity, and the choice of which can be made self consistently from the requirement that the scatterer and the medium together should not produce further scattering on the average [86,87]. The mean idea of the calculation [88] is as follows. Let

$$\langle G(E)\rangle = (E - H_{eff})^{-1} \tag{31}$$

and if $K = K(E)$ is an approximation of H_{eff}, than

$$\langle G\rangle = R + R(H_{eff} - K)\langle G\rangle \tag{32}$$

where

$$R = (E - K)^{-1} \tag{33}$$

is the "unperturbed" Green operator. Introducing the T matrix by

$$G = R + RTR \tag{34}$$

$$\langle G\rangle = R + R\langle T\rangle R$$

from (32) and (34)

$$H_{eff} = K + \langle T\rangle(1 + R\langle T\rangle)^{-1} . \tag{35}$$

This equation can be used in two ways:
1. as a successive approximation formula: the new K approximation of H_{eff} is calculated by inserting T corresponding the former K into (35), or

2. equation $\langle T(K) \rangle_{H_{eff}} = 0$ is used to determine K (because if $\langle T \rangle = 0$ then

The first method is usually applicable only if there is available some small parameter in K, like the concentration of a dilute alloy. It may be termed non-self-consitent.

The second method is self-consistent and resolves the difficulties of the first, but it is less simple from mathemathical point of view.

The first way above was chosen by M. Inoue et al. (1975) [89] who presented a self consistent cluster theory. They placed a cluster in the origin and took into account the multiple scattering by the atoms in the cluster accurately while the other clusters whose distribution was considered as completely random were taken as an average medium in the sense of the CPA. In a next paper (S. Yoshino [90], 1975) they applied their theory for 1, 2 and 8 atom clusters of amorphous Si. The structures of the clusters were chosen as follows: the configuration of atoms in 2-atom cluster is that of the nearest neighbour pair in the crystal, and by adding the six nearest neighbours atoms to this pair the 8 atom cluster (staggered configuration) is obtained. Thus

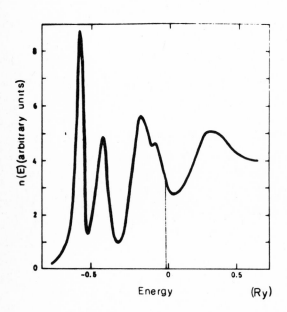

interatomic spacing in the cluster is equal to that of the crystal. These structures can be regarded approximately as the constituent parts of amorphous Si from radial distribution function data. Fig. 8 shows the local density of states of amorphous Si for 8 atom cluster. The Fermi energy is indicated by a dotted line. (Note the peak in the gap region is very similar to that on Fig. 6!)

Fig. 8.

The state density for the 8 atom cluster is obtained with global similarity to that of the band in the crystal and analysed from the bond formation by sp^3 hybridized orbitals. They succeeded to obtain a pseudogap at the Fermi energy in the density of states. In the 2 atom cluster case only slightly depressed density of states curve was found so it can be said that the pseudogap is made by the tetrahedral environment of four nearest neighbours and not by the formation of a pair of atoms.

This cluster-variation of CPA seems to be the most applicable to our problem.

However, we have to note, that the cluster type Green function methods mentioned at the point defect problem also does not utilize any long range order. Therefore these are suitable to any short range order problem. E.g. Jacobs [91] used the continued fraction method to the density of states of a disordered alloy with Bethe lattice. He used a second-quantitized Hamiltonian:

$$H = \sum_C \varepsilon_C a_C^+ a_C + \sum_{i \neq j} t_{ij} a^+ a_j \tag{36}$$

where i and j are site indices, a_j is the electron annihilation operator, $\varepsilon = \pm \frac{1}{2}\delta$ (the sign depending on whether the site is of type A or B), the second summation is taken over nearest neighbours only and t_{ij} takes the values t_{AA}, t_{BB}, t_{AB}, t_{BA} depending on the kind of atom in site i and j. In a Wannier representation then for e.g. $\varepsilon = 0$

$$\langle 0|G|0 \rangle = \cfrac{1}{E \,(0|G|0) - \cfrac{|(0|H|1)|^2}{E-(1|H|1) - \cfrac{|(1|H|2)^2}{E - \ldots}}} \tag{37}$$

where $|i\rangle$-s are members of a complete orthonormal basis formed from the $|0\rangle$ Wannier function of the site 0, applying the Hamiltonian (36) to $|0\rangle$ and using the Smith-orthogonalisation procedure. For the case $\varepsilon \neq 0$ one can use the idea of the CPA. However the extension of this method to clusters containing short range ordered units is still lacking and in the present single site approximation is too complex for calculations of practical importance.

SUMMARY

We compared the experimental properties of the different types of localized states according to Cheng with the calculated properties by various physical models suggested by several authors.

We stated that none of these models can account for all the localized states observed experimentally.

We suggested a natural way of classification for the electronic states in the Si/SiO_2 system based on the knowledge about the real structure of the system considering the oxide and the interface together as a disordered system.

We concluded that though the calculations made on short range ordered systems up till now are qualitative in nature, we can state that this qualitative picture given is correct in describing the states of the Si/SiO_2 system.

The distortion of the bonds between SiO_4 and Si_4 units in the interface, relative to those between two SiO_4 units causes a prolonged band tailing of both the valence and conduction band of SiO_2. These tails may correspond to the measured density of donor and acceptor interface states continuously increasing toward the bands. A prolonged influence of a large electric field at sufficiently high temperature can expand the interface region producing slow traps. Furthermore some of the bonds can break. Such dangling bonds, however do not effect the shape of the distribution of states [84].

This interpretation of the interface states explains   beside the amphoteric nature (neutral center model), and qualitative energy distribution, the orientation dependence also in a natural way. Eventually one can say that this corresponds to the "interface disorder" model and contains the excess silicon model too. Therefore it is in agreement with knowledge about dependence on oxidation condition and effect of annealing.

On the other hand the band bonding has less importance in the oxide, while in the meantime the vacancies can produce discrete levels with a distribution according to their relaxed environment.These levels can be responsible for oxide charges accumulating at the interface due to the mechanical stress of the interface, and can correspond to a part of the ionized traps far from the interface. This oxygen vacancy model or E_1' centre model agrees also well with observations.

Of course there can be some other centers in a real system, as impurities, dislocations etc. However these are not intrinsic to the system and can be avoided by suitable technology.

Finally we note, that the way of particular understanding of the problem is to find the special forms of these bond bending and bond breaking states via quantitative calculations.

REFERENCES

1. Y.C.Cheng, Progr. in Surf. Sci. 8, 181 (1977)
2. R.J.Kriegler, Y.C.Cheng, Dr.R.Colton, J.Electrochem. Soc. 119, 388 (1972)
3. F.Montillo, P.Balk, ibid. 118, 1463 (1971)
4. A.Goetzberger, E.Klausman, M.J.Schulz, CRC Crit. Rev. Solid State Sci. 6, 1 (1976)
5. R.A.Sigsbee, R.H.Wilson, Appl.Phys.Lett. 23, 541 (1973)
6. H.Deuling, E.Klausmann, A.Goetzberger, Solid State Electronics 15, 559 (1972)
7. H.Sakaki, K.Hoh, T.Sugano, IEEE Trans. El. Dev. ED-17, 892 (1970)
8. M.Kuhn, Solid State Electronics 13, 873 (1970)
9. P.V.Gray, D.H.Brown, Appl.Phys.Lett. 8, 31 (1966)
10. Y.Taru, K.Nagai, Y.Hayashi, Electronics Comm. Japan 53-C, 146 (1970)
11. M.Schulz, N.M.Johnson, Solid State Commun. 25, 481 (1978)
12. K.Ziegler, Appl.Phys.Lett. 32, 249 (1978)
13. E.M.Nicollian, A.Goetzberger, Bull. Syst. Techn. J. 46, 1055 (1967)
14. B.E.Deal et al., J.Electrochem. Soc. 114, 266 (1967)
15. D.R.Lamb, F.R.Badcock, Int. J. Electronics 24, 11 (1968)
16. J.M.Aither et al., IEEE NS-24, 2128 (1977)
17. T.H.Ning, Jap. J. Appl. Phys. 47, 1079 (1976)
18. M.H.Woods et al., ibid 47, 1082 (1976)
19. D.J.Di Maria et al., ibid. 47, 2740 (1976)
20. V.J.Kapoor et al., ibid. 48, 739 (1977)
21. R.Williams, Phys.Rev. 140, A569 (1965)
22. D.J.Di Maria et al., Phys.Rev. B11, 5023 (19)
23. J.M.Thomas et al., J.Phys.Chem.Sol. 33, 2197 (1972)
24. B.E.Deal, J.Electrochem.Soc. 121, 198 C (1974)
25a. T.W.Sigmon et al., Appl.Phys.Lett. 24, 105 (1974)
25b. W.L.Harrington et al., ibid 27, 644 (1975)
26. I.Bársony, Thesis (1977)
27. A.G.Révész, Phys. Stat. Sol. 24, 115 (1967)
28. N.Nagasima, Jap. J. Appl. Phys. 9, 879 (1970)
29. R.Williams, A.M.Goodman, Appl. Phys. Lett. 25, 531 (1974)
30. R.Flitsch, S.I.Raider, J.Vac.Sci. and Techn. 12, 305 (1975)
31. R.A.Clarke, D.L.Tapping et al., J.Electrochem.Soc. 112, 1347 (1975)
32. S.I.Raider, R.Flitsch, R.Rosenberg, Electrochem. Soc. Abstr. 136, 363 (1977)
33. J.S.Johanessen, W.E.Spicer, Y.Strausser, J. Appl. Phys. 47, 3028 (1976)
34. C.R.Halms et al., IEEE Trans. El. Dev. ED-24, 1208 (1977)
35. T.H.Di Stefano, J. Vac. Sci. and Techn. 13, 856 (1976)
36. O.L.Krivanek, T.T.Sheng, D.C.Tsui, Appl.Phys.Lett. 32, 437 (1978)
37. F.Stern, Solid State Commun. 21, 163 (1977)
38. K.H.Beekman, N.J.Harrick, J.Electrochem.Soc. 118, 614 (1972)
39. A.G.Révész, ibid 124, 1811 (1977)
40. A.Goetzberger, V.Heine, E.H.Nicollian, Appl. Phys. Lett. 12, 95 (1968)
41. A.G.Révész, K.H.Zaininger, R.J.Evans, J.Phys.Chem.Sol. 28, 197 (1967)
42. C.F.Neumark, Phys.Rev. B1, 2613 (1970)
43. Y.C.Cheng, Surf. Sci. 23, 432 (1970)
44. V.G.Litovchenko, Sov.Phys.-Semiconductors 6, 696 (1972)
45. T.W.Hickmott, J.Vac.Sci. and Techn. 9, 311 (1972)
46. T.Iizuka, T.Sugano, Jap. J. Appl. Phys. 12, 73 (1973)
47. A.G.Révész, IEEE Trans. El. Dev. ED-12, (1965)
48. C.M.Lane, IEEE Trans. El. Dev. ED-15, 998 (1968)
49. J.E.Thomas, D.R.Young, IBM J.Res.Develop. 8, 368 (1964)

50. F.M.Fowkes, T.E.Burgess, Electrochem. Soc. Abstr. 112, 261 (1969)
51. Y.P.Kwon, ibid. 109, 282 (1970)
52. A.Sólyom, Thesis (1978)
53a. I.Krafcsik, D.Marton, 1st Int. Conf. Surf. Sci., Amsterdam (1978)
53b. I.Krafcsik, D.Marton, Phys.Letters 71A, 245 (1979)
54. K.L.Yip, W.B.Fowler, Phys.Rev. B11, 2327 (1975)
55. F.Yndurian, J.Rubio, Phys.Rev.Lett. 26, 138 (1971)
56. N.F.Mott, Int.Conf. on Phys. of Semicond., Rome (1975)
57. R.C.Hughes, Applied Phys. Letters 26, 436 (1975)
58. S.T.Pantelides, J.Vac.Sci. and Techn. 14, 965 (1977)
59. S.T.Pantelides et al., Solid State Commun. 21, 1003 (1977)
60. S.T.Pantelides, W.A.Harrison, Phys.Rev. B13, 2667 (1976)
61. E.Calabrese, W.B.Fowler, ibid. 18, 2888 (1978)
62. Ph.M.Schneider, W.B.Fowler, ibid. 18, 7122 (1978)
63. J.Giber, P.Deák, D.Marton, phys.stat.sol.(b) K89 (1977)
64. S.T.Pantelides, Rev.Mod.Phys. 50, 797 (1978)
65. J.D.Levine, Phys.Rev. 140, A586 (1965)
66. G.F.Koster, J.C.Slater, Phys.Rev. 95, 1167 (1954)
67. J.Bernholc, S.T.Pantelides, Phys.Rev. B18, 1780 (1978)
68. S.T.Pantelides, Phys. of Semicond. 1978, Inst. Phys. Conf.
    Ser. No.43
69. M.Jaros, ibid.
70. F.Bassani, G.Iadonisi, B.Preciosi, Phys.Rev. 186, 735 (1969)
71. E.Kauffer, P.Pecheur, M.Gerl, Phys.Rev. B15, 4107 (1977)
72. J.D.Joannopoulos, E.J.Mele, Solid State Commun. 20, 729 (1976)
73. S.G.Louie et al., Phys.Rev. B13, 1654 (1976)
74. R.A.Evarestov et al., Phys.Stat.Sol. 79, 743 (1977)
75. D.J.Chadi, M.L.Cohen, Phys.Rev. B8, 5747 (1973)
76. P.Deák, next paper in this issue
77. A.J.Bennett, L.M.Roth, J.Phys.Chem.Solids 32, 1251 (1971)
78. K.C.Pandey, J.C.Phillips, Phys.Rev. B13, 750 (1976)
79. J.A.Appelbaum, D.R.Hamman, Phys.Rev. 136, 2166 (1972)
80. G.A.Baraff, J.A.Appelbaum, D.R.Hamman, Phys.Rev. B14, 588 (1976)
81. E.Louis, Solid State Commun. 24, 849 (1977)
82. S.F.Edwards, Phylos. Mag. 6, 617 (1961)
83. J.L.Beeby, S.F.Edwards, Proc. Roy. Soc. (London) A274, 395 (1962)
84. J.Keller, J.Phys.C. 4, 3143 (1971)
85. J.Keller, R.Jones, J.Phys.F: Metal Phys. 1, L33 (1971)
86. M.Lax, Rev.Mod.Phys. 23, 297 (1951)
87. P.Soven, Phys.Rev. 156, 809 (1967)
88. B.Velicky, S.Kirkpatric, H.Ehrenreich, Phys.Rev. 175, 747 (1968)
89. M.Inoue, S.Yoshino, J.Okazaki, J.Phys.Soc.Japan 39, 780 (1975)
90. S.Yoshino, M.Inoue, M.Okazaki, J.Phys.Soc.Japan 39, 787 (1975)
91. R.L.Jacobs, J.Phys.F. 4, 1351 (1974)
92. D.J.Chadi and M.L.Cohen, phys.stat.sol.(b) 68, 405 (1975)
93. I.Bársony, D.Marton, J.Giber, Thin Solid Films 51, 275 (1978)

CYCLIC CLUSTER MODEL (CCM) IN THE CNDO APPROXIMATION
FOR DEEP LEVELS IN COVALENT SOLIDS

P. Deák
Physical Institute of the Technical University Budapest
H-1502 Budapest, P.O.Box 112, Hungary

This contributed paper joins prof Giber's invited paper at the point of deep level calculation problems. I'm in the convenient situation to omit habitual summary of the literature in the topic, because I can refer the review-paper of Pantelides [1] mentioned already in the previous lecture (Fig. 1).

Instead of the particular discussion of the different methods, what could be the subject of a longer paper only, I should like to emphasize that all the methods intended to quantitative calculations are extremely laborious and computertime-consuming. This is the case even for an ideal vacancy, too.

For that reason I should like to turn the attention to a possibility, which was being omitted from the classification of Pantelides. This method can be placed in Fig. 1 under the label of supercell methods. It is an improvement of the MUCA (Molecular Unit Cell Approximation) of Messmer and Watkins [2], toward self-consistency. However, it remains in the LCAO-MO scheme and so it is much simpler than the self-consistent pseudopotential calculation of Louie and coworkers [3].

In order to introduce it, let's turn to the outline of supercell methods in the LCAO-MO picture. The idea of supercell methods means a bandstructure calculation, which employs the multiple of the elementary cell. This large unit cell contains the defect, e.g. a vacancy. Of course, that is also repeated with the periodicity of the large unit cell.

In the case of a band structure calculation the Roothaan-equations turn to the equations of the crystalline orbital method [4].

$$\sum_B \sum_\nu \left[F_{\mu\nu}(\underline{k}) - \varepsilon_n(\underline{k}) S_{\mu\nu}(\underline{k}) \right] C_{\nu n}^B(\underline{k}) = 0$$

where $A,B,C,\ldots$; $i,j,k,\ldots$; $\mu,\nu,\lambda,\ldots$; $m,n,\ldots$ etc. label atomic sites, different large unit cells, atomic orbitals and crystalline orbitals respectively. That is μ_i^A represents the μ-th atomic orbital centered on the A-th atomic site in the i-th unit cell. Than the Fock matrix is:

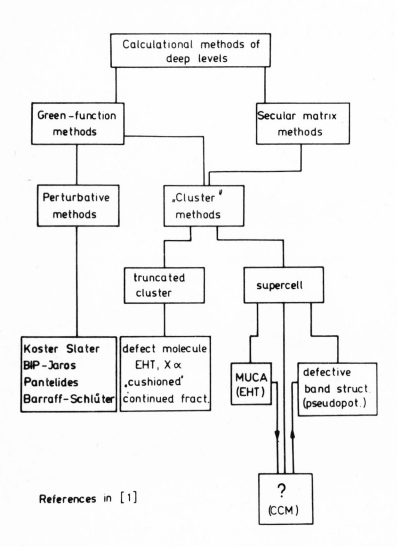

Fig. 1.

$$F_{\mu\nu}(\underline{k}) = \sum_{j}^{N} F_{\mu\nu}^{oj} e^{i\underline{k}\underline{R}_j}$$

$$F_{\mu\nu}^{oj} = H_{\mu\nu}^{oj} + \sum_{\underline{k}'} \sum_{C,D} \sum_{\lambda\sigma} \frac{P_{\lambda\sigma}(\underline{k}')}{N} \sum_{k,l}^{N} \left[(\mu_o^A \nu_j^B | \lambda_k^C \sigma_l^D) - \right.$$

$$\left. - \frac{1}{2}(\mu_o^A \lambda_k^C | \nu_j^B \sigma_l^D) \right] \cdot e^{i\underline{k}'(\underline{R}_k - \underline{R}_l)}$$

$$(\mu\nu|\lambda\sigma) = \int \mu\nu \frac{1}{r_{12}} \lambda\sigma \, dr^3$$

$$P_{\lambda\sigma}(\underline{k}') = 2 \sum_{m}^{occ.} C_{\lambda m}^{C*} C_{\sigma m}^{D}$$

$$S_{\mu\nu}(\underline{k}) = \sum_{j} S_{\mu\nu}^{oj} e^{i\underline{k}\underline{R}_j} = \sum (\mu_o^A / \nu_j^B) e^{i\underline{k}\underline{R}_j}$$

$$(\mu/\nu) = \int \mu\nu \, dr^3$$

Where $H_{\mu\nu}^{oj}$ is an element of the core Hamiltonian matrix, $P_{\lambda\sigma}(\underline{k}')$ and $S_{\mu\nu}(\underline{k})$ are the bondorder and overlap matrices respectively.

$C_{\nu\eta}^{B}$ -s are the coefficients of the linear combination.

The Fock matrix of a $\underline{k}$ state depends on all other $\underline{k}'$ states. This coupling, i.e. the summation over $\underline{k}'$ in F, and further the summation over λ and σ (many-center integrals) and that over the cells (infinite range of interactions) makes the calculation in an *ab initio* manner hopeless.

This problem is avoided by the application of the EHT (in MUCA) [2] which neglects the second term in the Fock matrix and parametrizes the remaining part semiempirically. However, the price of that is high. The neglection of electron-electron interactions makes the application to relaxation problems unreliable, while the EHT, which is parametrized for ground state properties, doesn't give a proper conduction band.

The way we chose was the introduction of the CNDO (Complete Neglect of Differential Overlap) method. This is a well tested self-consistent quantumchemical method with parametrizations for conformation analysis (CNDO/2) [5], for ionization and excitation energies (CNDO/S) [6], and for heats of formation (CNDO/SW) [7]. The CNDO maintains a part of the electron-electron integrals, but evaluates these approximately. The different orbitals on the same atom is handled uniformly, so the sum over orbitals in the Fock matrix disappears.

Of course, the coupling between $\underline{k}$-states returns. Therefore, let's make an other restriction: we'll carry out calculations for the $\underline{k} = 0$ state only, and will neglect the interaction of this to all other $\underline{k}'$-states. We note however, that the defectless large unit cell contains a set of representative points of the original Brillouin-zone, and the interactions between these are preserved [8].

In this way we can avoid the summation over the Brillouin-zone. A further simplification can be introduced by the limitation of the summation over different cells, in order of avoiding direct defect--defect interactions. More exactly, we'll employ a cut-off distance of the interactions, so that they will be just shorter than the linear cluster size. This will cause the lattice sums to disappear from the Fock-matrix.

All these restrictions mean, that instead of a band structure calculation we'll performe a molecular cluster calculation with periodic boundary conditions. Of course, these simplifications bring back the convergence problems characteristic to the cluster model. These are:
1. Convergence with *defect-defect separation* for excluding defect--defect interactions.
2. Convergence with cluster size, i.e. *how many* original $\underline{k}$-*state* must be *included* in the large unit cell to reproduce the perfect crystal band structure.

3. Convergence with *cut-off distance of interactions*, i.e. how many
   shell of neighbours must be included.

   Regarding the last one, we refer to the result of A.Zunger [9].
He carried out two parallel calculations. In the first he employed a
self-consistent Fock-matrix over 180 $\underline{k}$-points, while in the second
he applied the Fock-matrix of a limited size cluster. As it can be
seen the results show only a little difference (Table I). Now, for our
case, we repeated the calculation of Evarestov and coworkers on a C-16
cluster of diamond with CNDO/2 parametrization [10], but without calculation lattice sums (Fig. 2). As it can be seen the results here are
also similar.

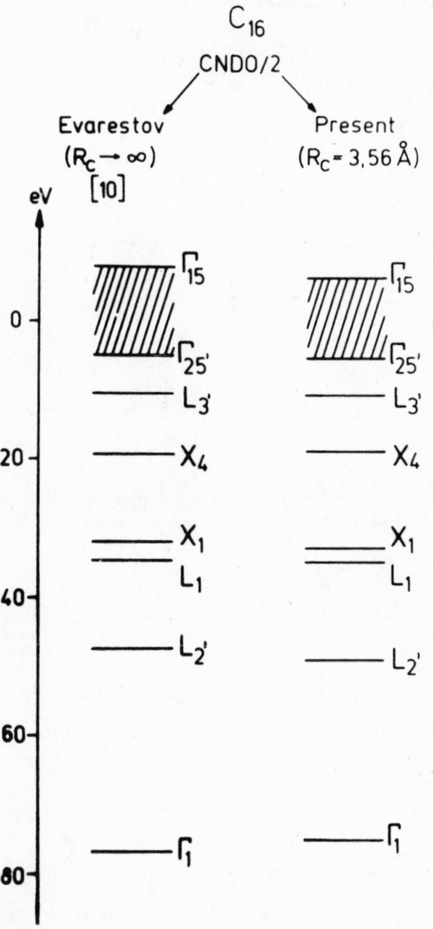

Table I.

	SCF	Cluster
gap	13.3	12.8
π band width	21.3	20.9
total band width	47.8	46.9

Fig. 2.

Regarding the convergence
with included $\underline{k}$-states, we refer
again to Evarestov, who showed
that for LiF the application of
54 atoms instead of sixteen alters
the charge distribution in a range
less then 10 %. That means, that
the result on a set of the
$\{\Gamma,X,L,\Delta,\Sigma,W\}$ points differs
only slightly from that of
$\{\Gamma,X,L\}$. Perhaps for the case of
pure covalent materials the situation is not so favourable, but
a strict analysis stands out of
our computer possibilities.
Therefore the only one we can do
is the comparison of our results
directly with experiments. The
results on the C-16 cluster by
the CNDO/S method reparametrized
for tetrahedral molecules can be
seen on Fig. 3. We note that be-

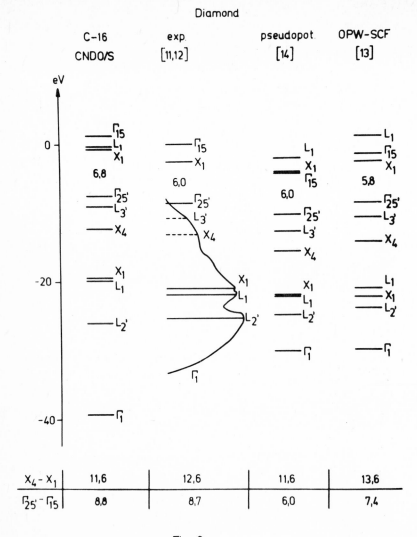

Fig. 3.

cause of the calculation is made in a molecular picture, we might not except the conduction band to agree with the virtual orbitals of the system. That is because the separation of the conduction and the valence bands are known from optical transitions, for which the Koopman's theorem does not hold any more, contrary the case of ioniz-ation energies. However, the CNDO/S allows us to calculate the tran-sition energies in a self-consistent manner including a limited CI cal-culation for correlation effects.

On the picture one can see the comparison of our results with experiments [11,12] and with band structure calculations [13,14]. We emphasize, that the agreement is achieved not by fitting, but by optimalization of the parameters for small hydrocarbons. On the other hand a standard CNDO/2 calculation yields the equilibrium lattice constant also very accurately (3.54 to 3.56 Å).

The only large disagreement is in the position of Γ_1 and correspondingly of Γ_2, which latter isn't shown on the figure. The cause of this is an inherent feature of the CNDO, namely keeping the exponents of s- and p-orbitals the same.

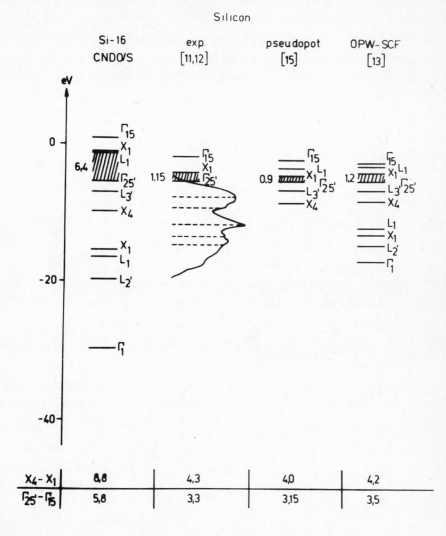

Fig. 4.

Unfortunately we couldn't achieve such good results for the silicon conduction band (Fig. 4). The reason of this is that we are unable to involve d-orbitals, which are essential in calculating excited-state properties of third-row elements. However, we note the agreement of the valence band and especially that of the first ionization potential with experiments.

However, it's hard to say anything for the convergence with defect-defect separation. Louie and his coworkers [3] found that a 54 atom cluster is required to overcome this problem. Unfortunately this is out of our computer capacity. However, as experimental data show [18], the 65 % of the vacancy wavefunction is localized on the four nearest neighbours and only the remaining term is diffuse. Therefore we hope, that the cutting off of the interaction distance omits the main part of the defect-defect interaction without significant decrease in the quality of results. This hope is supported by the results of all previous calculations on the change in total electronic density, which is almost fully restricted to the area of the vacancy.

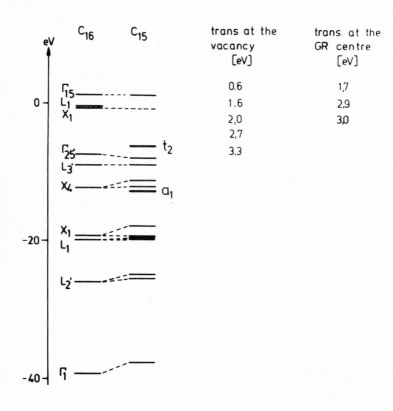

Fig. 5.

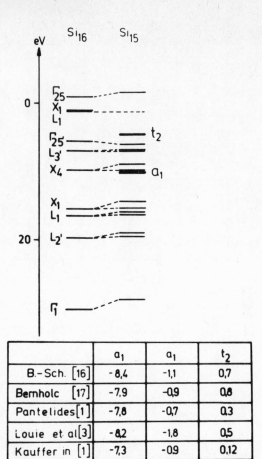

	a_1	a_1	t_2
B.-Sch. [16]	-8,4	-1,1	0,7
Bernholc [17]	-7,9	-0,9	0,8
Pantelides[1]	-7,8	-0,7	0,3
Louie et al[3]	-8,2	-1,8	0,5
Kauffer in [1]	-7,3	-0,9	0,12

Fig. 6.

Unfortunately, we can't compare our preliminary results (Fig. 5) on vacancies directly with experiments. The reason of this is that the partially filled t_2 state in the gap causes a strong Jahn-Teller distortion, which isn't involved in our calculations yet. (Calculations including this effect are under work). Furthermore, in the case of silicon, the role of the conduction band in creating the t_2 state cannot be neglected and so the results for silicon can be regarded as qualitative ones only (Fig. 6). On the other hand, no such expanded calculations are available for diamond vacancy as for that in silicon. Anyway the results are qualitatively correct, as it can be seen by comparing the electronic transitions at the diamond vacancy with experiments. We have to note, however, that the too high position of the reasonant a_1 level indicates perhaps the necessity of a larger cluster.

By all means the procedure is much simpler than those comparable with it in efficiency. This has a particular importance from the view-point of materials with larger elementary cell, e.g. SiO_2.

Unfortunately I can't give account of calculations on this material, because the d-orbital problem is still under work.

REFERENCES

1. S.T.Pantelides, Rev.Mod.Phys. 50, 797 (1978)
2. R.P.Messmer, G.D.Watkins, in "Radiation Damage in Semiconductors"
   Inst. Phys. Conf. Ser. No.16, 255 (1973)
3. S.G.Louie, M.Schlüter, J.R.Chelikowsky, M.L.Cohen, Phys.Rev. B13,
   1654 (1976)
4. J.-M.André, L.Gouverneur, G.Leroy, Int.J.Quantum Chem. 1, 427,
   451 (1967)
5. J.A.Pople, D.P.Santry, G.A.Segal, J.Chem.Phys. 43, S129 (1965);
   44, 3189 (1966)
6. J.Del Bene, H.H.Jaffé, J.Chem.Phys. 48, 1807 (1968); 50, 1126
   (1969)
7. J.M.Sichel, M.A.Whitehead, Theoret.Chim.Acta 7, 32 (1967);
   11, 220,239,254 (1968)
8. R.A.Evarestov, Phys.Stat.Sol.(b) 72, 596 (1975)
9. A.Zunger, Phys.Rev. B13, 5560 (1976)
10. R.A.Evarestov, V.A.Lovchikov, Phys.Stat.Sol.(b) 79, 743 (1977)
11. R.G.Cavell, S.P.Kowalczyk, L.Ley, R.A.Pollak, B.Mills,
    D.A.Shirley, W.Perry, Phys.Rev. B7, 5313 (1973)
12. J.C.Phillips, Solid State Phys. 18, 55 (1966)
13. F.Herman, R.L.Kortum, C.D.Kuglin, Int.J.Quantum Chem. 1S, 533
    (1967)
14. G.S.Painter, D.E.Ellis, A.R.Lubinsky, Phys.Rev. B4, 3610 (1971)
15. D.Brust, Phys.Rev. 134A, 1337 (1964)
16. G.A.Baraff, M.Schlüter, in "Semiconductor Physics" Inst. Phys.
    Conf. Ser. No.43, 425 (1979)
17. J.Bernholc, S.T.Pantelides, N.O.Lipari, ibid. No.43, 429 (1979)
18. G.D.Watkins, Chinese J. Phys. 15, 92 (1977)

IS THERE A MINIMUM LINEWIDTH IN INTEGRATED CIRCUITS?

J. Torkel Wollmark
Research Laboratory of Electronics
Chalmers University of Technology

402 20 Göteborg 5 - Sweden

ABSTRACT

Reduction of linewidth is one of the most important problems in integrated circuit technology. The progress made is reviewed and the question whether there exists a minimum linewidth set by physical effects is discussed. It is suggested that two fundamental criteria exist which set such a limit. First, a minimum spot size, 2-3 nm, of the fabricating beam is determined by the Heisenberg uncertainity principle, by the minimum particle size in the resist, and by particle scattering. Second, statistics of pattern delinaation with the accompanying probability of an occasional deviation larger than a set tolerance limit sets a linewidth limit in the range 30-70 nm depending on the complexity of the circuit pattern. Other factors which enter in particular cases are also mentioned.

INTRODUCTION

The future of integrated circuits depends strongly on the success of the continuous struggle to further miniaturize the circuit patterns. There are several reasons for this. The most important reason has to do with the fact that the substrate used for silicon monolithic circuits contins on their surface a sensity of statistically distributed flaws, each one potentially fatal to a circuit containing the flaw. Therefore the best strategy has been to miniaturize the circuit pattern until many patterns fall in between the flaws resulting in a reasonable yield of perfect circuits and consequently a reasonable cost per circuit. It is easy to see that as the linewidth decreases the yield increases much faster than linearly.

Another good reason for miniaturization has been small size and high packing density of the finished circuit. This is desirable in for example magnetic circuits such as bubble memories which compete with disk or drum memories which have a very high packing density of bits in the magnetic medium. Small size is also desirable for silicon integrated circuits so that they may fit into existing and thoroughly tested hermetic packages. On a higher level the cost of a system is much reduced if the entire electronics of the system may be places on one chip and in one single package.

The flaws on the surface of the silicon substrates come from two groups of causes, one originating from material parameters, and one from handling. Material parameters introducing flaws are mechanical strain set up by thermal expansion differences between the silicon crystal and the covering oxide and metal layers, and strain introduced by the necessary doping atoms which in some areas reach high concentrations. As the strain is high it is relieved by forming crystal and particularly oxide faults in some spots going as far as microcracks. These microcracks serve as preferred deposit regions for impurities during the various processing stages and as preferred attack points for etches. The other group of defects comes from the handling of the wafers during processing. The wafers must be handled mechanically by tools like tweezers and are stacked or swirled in beakers during washing. Each time a mechanical stress adds to whatever stress is already present. The handling during processing also means temperature cycling, again adding to the stress. Each processing step adds new impurities, mostly unwanted, and not the least as a result of washing in impure solvents and handling in "clean" air. The resulting flaws in the insulator and metal layers on the silicon surface are sometimes discernable electrically or optically but most often defy observation and remain hidden perhaps to turn up at later stage during repeated temperature cycling or repeated electrical stress during the life of the circuits.

Can the flaws be avoided or are they fundamental? The answer to this question is not yet clear. Obviously many types of flaws can be avoided by improved perfection in handling and it is well known that disciplin and care in processing has a considerable influence on the yield in fabrication. However, some of the flaws, particularly those originating from the thermal expansion differences and lattice mismatch, appear fundamental or at least not easily eliminated in a near future. Also, economy of processing always will favour a certain laxity in handling and consequently also a certain density of flaws. Based on this argument the effort to reduce linewidth will continue.

Considerable progress in miniaturization has been made over the years. The "small" dimensions of circuits in the early 1960s with lines of 25 μm width have evolved in the circuits of today into lines of about 5 μm width. In some very complex circuits linewidths of 3-4 μm are appearing. In the laboratory and for not so complex circuits (bubble memories), linewidths down to 0.1-0.3 μm have been reported. A natural question is then: How much further reduction of linewidth is possible?

In practice, there are a great many factors that may influence the minimum line width - spacecharge spread of electron beams, chromatic

aberration of optical or electron-optical lenses, beam deflection distortion, electron initial velocity spread, vacuum insufficiencies,
metal migration etc.

The exact treatment of all these effects is very difficult and
depends on many assumptions which are usually not fundamental. The
analysis thus hinges on a wide collection of "ifs". On the other hand
a fundamental lower limit on line-width may be based on four factors
only - the Heisenberg uncertainity principle, the photoresists molecular
size and particle scattering (which together determine the minimum
spot size of the fabricating beam) and finally the statistics of
pattern formation. These factors are simple and represent unavoidable
physical limitations which need very few assumptions. It is believed
that the constitute a fundamental basis and that many of the other
effects mentioned above may present additional complications which
raise the minimum line-width somewhat. The fact that the minimum line
 width obtained in the laboratory is very close to the limit derived
here suggests that with proper optimisation additional factors may be
looked upon as creating second-order effects.

THE MINIMUM SPOT SIZE IN INTEGRATED CIRCUIT PATTERNS

In order to transfer the information about a circuit into a layout of the circuit pattern, a tool is needed. The ultimate tool for
fine resolution is a beam of particles - photons, electrons, ions etc.
Until recently all pattern fabrication was made by photons - light
beams, with an energy of a few electron volts (a wavelength λ in the
ultraviolet of 200-450 nm). As is well known, this limits the minimum
spot size to about a wavelength of the light used, or about 200 nm
(2.000 $\overset{o}{A}$).

The fundamental reason for this limit rests in the Heisenberg
uncertainty principle which states that the product of the uncertainties in location (ΔL) and in momentum (ΔP) must exceed Planck's constant h, or

$$\Delta L \Delta P \geq h \tag{1}$$

The momentum may vary at most from +p to -p. Then the minimum uncertainty L is

$$L \geq \frac{h}{2 p} \tag{1}$$

where p for photons is

$$p = h/\lambda \tag{2}$$

giving
$$L \geq \lambda/2 \tag{3}$$

or
$$L \geq hc/2E \tag{4}$$

where E is the energy of the beam, c is the velocity of light and λ is the wave length.

For electrons or ions of moderate energies the momentum p is

$$p = mv = (2mE)^{1/2} \tag{5}$$

giving
$$L \geq \frac{h}{2} (2mE)^{-1/2} \tag{6}$$

where m is the beam particle mass.

More recently, pattern layout work has utilised electron beams with electron energies of 10-30 keV (a wavelength much less than 0.1 μm, 1.000 Å). However, in that case the spot size is limited to about 1-10 nm, say 2.5 nm, because of the graininess of the photoresist and because of scattering of the electrons in the target as indicated in Fig. 1. The shaded area shows the excluded region of spot sizes.

The scatter region requires some comments. The scattering problem can be partially circumvented by using beams of high energy which penetrate the photoresist layer before much scattering has had time to develop and thereby retain a small beam spot. The scatter shower is then placed deep inside the substrate (e.g. silicon) where it may later be annealed and thereby made practically harmless. In line with this electron microscope studies of features with dimensions under a nano-metre have been made [1].

However, at the ultimately small line-width analysed here, con-trast in the fabricating process cannot be too low. Therefore scattering and backscattering reducing contrast cannot be tolerated, or alternatively will in-crease the minimum line-width. A qualitative impression of the amount of back-scattering may be obtained from Fig. 2 which shows results by Murata et.al. [5]. A quantitative treatment of this problem is complex and will not be attempted here.

The minimum spot size ΔL, besides limiting the smallest detail in the circuit pattern (say the smallest line-width) at the same time means that we do not know any mathematical point in the pattern area to better than ΔL. We may therefore speak of ΔL as the uncertainty of the beam in the meaning of the Heisenberg uncertainty principle. The

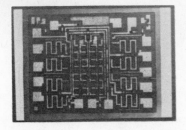 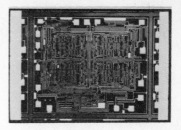

Fig. 1. Three consecutive stages spanning about 10 years in the evolution of integrated circuits. Each circuit has 10x more elements and 50 % reduced line-width compared to the previous one (Motorola Semiconductor Products Inc.)

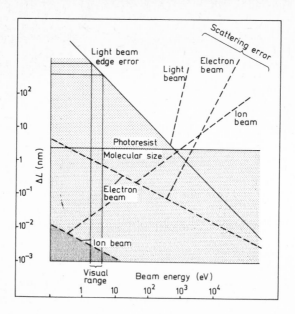

Fig. 2. Uncertainty in location of a beam particle vs particle energy. Shaded
regions are forbidden by the Heisenberg uncertainty principle (left-
sloping lines), the minimum molecule size of the photoresist (horizontal
line) and particle scattering at high energies (right-sloping lines)

meaning of the uncertainty is that the mathematical probability of
finding the beam (or more exaclty a beam particle), within the stated
uncertainty ΔL, is exactly 1/2 by definition.

THE MINIMUM DIMENSION IN INTEGRATED-CIRCUIT PATTERNS

Assuming a tool with the fundamental uncertainty ΔL in location
of a point we will now proceed to use this tool to define a pattern,
such as a mask pattern used in fabrication of an integrated circuit.
We will do this in steps encompassing higher and higher complexity. In
the end we will find that when it comes to defining a practical pattern
we will have to multiply the fundamental uncertainty ΔL by a factor
of 3-4 to obtain the uncertainty in the pattern, i.e. the uncertainty
in geometrical dimensions, and therefore also electrical dimensions of
the circuit elements.

From one to two dimensions

First consider an arbitrary length l_o, defined by two end points,
each characterised by an uncertainty ΔL. We may consider this uncertainty in a statistical sense as a standard deviation of a normally
distributed random variable. The resulting length uncertainty Δl_o,

assuming that the two points are independent, is obtained from

$$\Delta l_o = \left[(\Delta L)^2 + (\Delta L)^2 \right]^{1/2} = 2^{1/2}\,\Delta L. \tag{4}$$

However, the different lengths in an integrated circuit pattern must be set out from one (or more) reference point in order to place each element in correct relation to all other elements. This reference point also is not know better than the uncertainty ΔL. Thus, starting with the simple case of one dimension, setting out a length colinearly as indicated schematically in Fig. 4, we obtain an uncertainty in length Δl_1

$$\Delta l_1 = \left[(\Delta l_o)^2 + (\Delta l_o)^2 \right]^{1/2} = 2\Delta L. \tag{5}$$

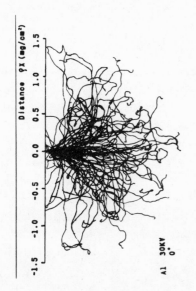

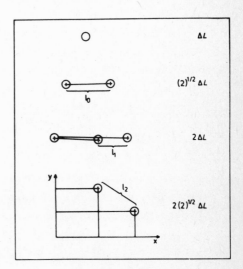

Fig. 3. Behaviour of incident electrons entering an aluminium target, until their energy of 30 keV is reduced to 0.5 keV in the target, calculated by the Monte Carlo method (from 5).

Fig. 4. Schematic diagram of the derivation of the uncertainty of an arbitrary circuit element dimension

The general case of a length in two dimensions with end points (x_1, y_1 and x_2, y_2) may be obtained in a two-step process, first setting out the x length, then the y length from, in this case, two perpendicular reference axes, which are again characterised by the uncertainty ΔL. The resulting uncertainty Δl_2 is obtained from

$$\Delta l_2 = \left[(\Delta l_1)^2 + (\Delta l_1)^2 \right]^{1/2} = 2(2)^{1/2} \Delta L. \qquad (6)$$

The meaning of the uncertainties derived above is the same as that for
a point, that the probability of a measurement being correct (i.e.
that the intended dimension is contained within the stated uncertainties) is exactly 1/2.

Proximity effects

As is well known from fabrication of integrated circuits, complications arise when circuit features are placed too close together;
there are so-called *proximity effects*. In a typical case the spacing
between two adjacent lines may grow together as the photoresist
receives exposure contributions by diffraction and reflection of light
from both lines.

A fundamental minimum proximity effect, in an otherwise ideal
case with negligible reflection, may be based on the uncertainty
principle and a normal distribution of exposing events in the photoresist. I will consider this for a one-dimensional case, i.e. for a
line structure.

Consider the fabrication of a line of minimum width d_{min} in which
the probability P of a fabricating beam particle (for example, for
exposing photoresist) hitting the target at a distance x from the line
centre obeys a normal distribution. The normal distribution means that
a finite (even if small) probability exists of hitting also a neighbouring line. In order to reduce this probability to negligible pro-
portions, the lines must be separated by a factor which may be computed
from probability theory to be 1.63.

Summing up the results, we have found that a minimum uncertainty
ΔL in the fabricating beam leads to a minimum uncertainty Δl in
circuit pattern dimensions of $\Delta l = 2x(2)^{1/2} x 1.63 x \Delta L = 4.6 \, \Delta L$. With
$\Delta L = 2.5$ nm we obtain a *dimension uncertainty of about 10 nm* for any
dimension in an integrated-circuit pattern.

MULTIPLE-ELEMENT CIRCUIT PATTERNS

Let us now consider a statistical model for plotting a circuit
pattern.

Random pattern

Assume as before a beam of particles with an uncertainty ΔL and the resulting uncertainty in each measurement Δl. Assume that the uncertainties are random and distributed with a normal distribution. Assume that the pattern requires N independent measurements. Assume further that proper functioning of the circuit does not allow any of the N measurements to deviate from the intended value by more than a tolerance of $\pm p$ %, or a maximum fractional deviation $\epsilon = p/100$. This latter assumption may need some clarification.

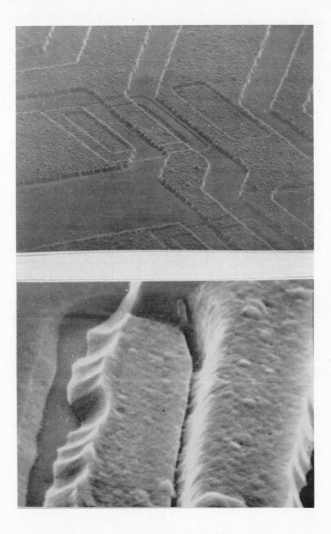

Fig. 5. a) magnified (500x) portion of an integrated circuit. Line-width is 10 μm.
  b) magnified (5.000x) detail showing structure of aluminium metallisation
     (centre and right) and oxide window (left)

Figure 5 shows a part of a typical circuit pattern. It consists of closely packed connection lines and circuit elements in the form of diffused regions - transistor-base regions, transistor-emitter regions, contact holes, etc. If the location of one edge of a connection line, for example, should deviate side-ways by an amound equal to the line--width, it would either cause an open line (line-width zero), or a shorted line (touching the neighbouring line). In other words, proper functioning of the circuit requires that everywhere $p < 100$ % or $\epsilon < 1$. Similar arguments apply to diffusion edges, contact holes, etc.

We will not apply a statistical approach. Consider as a random variable the deviation x of the location of the edge from the mean location, which in this case is the intended location. We will assume that each measurement of x is independent of each other measurement. We are interested in what happens when x reaches ϵd, i.e. when the deviation amounts to a prescribed tolerance ϵ of d, the minimum dimension, (e.g. the minimum line width). We then rewrite the random variable x in the form of a standardised random variable y by choosing

$$y = \frac{\epsilon d}{\sigma} \qquad (7)$$

where σ is the standard deviation, which is related to the minimum uncertainity Δl afflicting each measurement as shown below.

From probability theory the probability P that the edge location is within $y_1 = \epsilon_1 d_1$ from the mean location in a particular measurement, either on the left-hand or on the right-hand side, is given by

$$P = \frac{2}{(2\pi)^{1/2}} \int_{0}^{y1} \exp(-y^2/2)\, dy. \qquad (8)$$

The probability S that the location is outside the tolerance limit in an individual measurement is

$$S = 1-P. \qquad (9)$$

If the circuit requires N such measurements (N is approximately the number of resolvable picture points in the pattern area, we may think of N as in a loose sense approximately equal to the number of elements in the circuit), then the probability Y of a functioning circuit is

$$Y = (1-S)^N = p^N. \qquad (10)$$

In order to obtain a reasonable yield of good circuits, let us set

S = 1/(N+1). For N large, S ~ 1/N. This means that if the circuit requires 1000 measurements (approx. 1000 elements) the shrinkage S is required to be ≤ 0.001 per element.

To find the standard deviation σ, characterising the normal distribution, set N = 1, when P = S = 0.5, є = 1 and d = 10 nm. From eqn. 8, y_1 = 0.675, and from eqn. 7, σ = 14.8 nm.

Summarising this, we have:

$$d_{min} = \frac{y}{є} \times 1.48 \times 2(2)^{1/2} \times 1.63 \, \Delta L, \qquad (11)$$

where y is the value of a normalized variable appearing in a normal distribution with a probability of at most $(N + 1)^{-1}$.

Repetitive circuit pattern

In practical integrated circuits a particular measurement defining a detail in the circuit pattern, such as an emitter side or a contact hole edge, is in general not independent of other measurements. The dependence of measurements on other measurements enters the derivation of minimum line-width in a way that may be understood as follows.

Let us consider a circuit in the form of an element network with N elements, all nominally identical. Let an element be completely described by C measurements. We will assume the material ideal, with constant uniform doping, etc. We may then fabricate the circuit by first making one element and then, by what is usually called step-and--repeat processes, copy the element. We now have two elements. A further copying produces four, etc. The same sequence may of course be applied to the connection lines.

The necessary number of measurements B may be written as

$$B = C \cdot {}^2\log N + n, \qquad (12)$$

where N is the number of elements in the circuit. $z = {}^2\log N$ is the number of step-and-repeat processes ($N = 2^z$), C is the necessary number of measurements for one element, and n is the number of additional measurements (if any) needed for non-repetitive parts.

For $N = 10^5$, C = 3 and n = 0, we get B = 49.8. In other words, complete repetitiveness reduces B by a factor of about 10^3. For a conventional photographic mask C = 2 and B = 33.2. Once more it must be said that this represents an extreme case. In practical memory circuit, which are among the most repetitive circuits made, the minimum value of B above would be increased as n ≠ 0 because of geometry re-

quirements, input circuits, output circuits, battery leads, x- and
y-decoders, bonding pads, test patterns, etc. For that reason the
value of B would be intermediate between its value in a truly repetitive
circuit and its value in a truly non-repetitive ciruit. A representative value of n may be obtained by counting the necessary measurements
for the pattern.

This analysis may be extended to other forms of repetitiveness,
such as repeated use of the same transistor in the circuit or in circuit
parts, repeated use of line segments, etc. Any pattern part that may
be kept in a "library" for reuse means repetitiveness and may be
analysed as above. For any part taken from the "library" only two measurements are needed, corresponding to the location in the pattern of
two corners of the part.

The ratio R of the number of measurements necessary in repetitive
and non-repetitive circuits becomes

$$R = \frac{^2\log N}{N} + \frac{n}{CN}.$$ (13)

For a large circuit, with say 10^5 elements, the first term is of order
10^{-4} while the second term may be 10^{-1} or 10^{-2}. Then

$$R \approx n/CN$$ (14)

which means that the information content in the repetitive part may be
neglected compared to the non-repetitive part.

CONSEQUENCES OF THE PHYSICAL-STATISTICAL MODEL

Let us now consider the ramifications of the physical-statistical
model described, and what practical consequences in real integrated
circuits it may lead to Fig. 6 shows the minimum possible line-width
d_{min} versus the necessary *number of measurements* N, with the tolerance
ε as parameter, obtained from eqns. (7)-(10). We may think of N as
roughly the number of elements in the circuit if we assume that only
one measurement is needed for each element. As will be explained below
the minimum number of measurements per element may be in the range
2-10 for very simple standard elements. As a factor of 2-10 in N makes
only a modest difference in d_{min} in Fig. 6 we are justified in iden-
tifying N approximately with the number of elements in the circuit. A
more exact relation between N and the number of elements in the circuit
is obtained by counting the necessary minimum number of measurements.

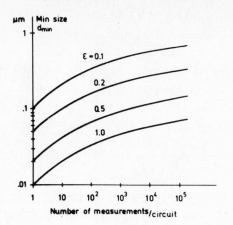

Fig. 6. Minimum line-width (d_{min}) vs number of measurements (N) necessary for fabricating the circuit pattern. Parameter is the allowed tolerance in line-width, ε .

As may be s en in Fig. 6, the minimum line-width increases from 10 nm to about 70 nm when N increases from 1 to 10^5. The change is rapid at low N but more gradual at large N, so that even the circuits of tomorrow with $N = 10^5-10^7$ will have very nearly the same d_{min}.

The influcence of *tolerances* (ε) is also apparent in Fig. 6. Tolerances of $\pm$100 % in line-width correspond to $\varepsilon = 1$. One should remember that -100 % in line-width represents an open line, while +100 % represents a doubling of the line-width, with possible short-circuit to an adjacent line, assuming line-width an line-spacing equal.

Tolerances in dimensions are related to tolerances in electrical characteristics of circuit elements such as transistors, resistors, etc. Doubling the emitter side in a transistor doubles the emitter current and effects indirectly many transistor paramters such as emitter impedance, current-transfer ratio, cut-off frequency, etc.

Which tolerance value is more realistic depends to a large extent on the circuit design, on the power level of the circuit, on circuit cost, etc. For a comparison, some values of ε representing todays' more typical values of 0.1 or 0.2 are shown in Fig. 6. A reduction of ε by a factor of 10 is balanced by an increase in d_{min} by the same factor.

DISCUSSION

All through this analysis the emphasis has been on limitations set by circuit connection lines rather than on limitations set by circuit elements. A good reason for this is that a large variety of circuit elements exist with very different properties and therefore also limitations. Circuit connection lines on the other hand are nearly identical for all types of integrated circuits, whether semiconductor, ferrite (bubble memories), superconductor, or other. Another good reason is that for complex circuits with many elements connected the line characteristics become dominating. However, in the ultimate analysis, elements and lines must be considered together.

Integrated circuits are always three-dimensional, even though only two dimensions have been treated in this paper. The reason for considering only two dimensions is partly (i) that much of the interest centres today around the problem of maskmaking in integrated circuits, a mainly two-dimensional problem, and partly (ii) that the third dimension, perpendicular to the surface, appears very different from the other two. Actually the limitations analysed in this paper enter in all three dimensions equally.

The smallest line-width reported today for practical circuits [3] is 100-300 nm which was obtained in bubble memory circuits the geometrically simplest and most repetitive circuits today. The molecule size (corresponding to ΔL in the analysis) in the electron resist used is not known but the molecules are usually long and threadlike. For that reason the minimum line-width derived in this article is very close to that actually observed.

REFERENCES

1. Brewer G.R. (1971) IEEE Spectrum 3 (January) 23-27
2. Brillouin L. (1956) Science and Information Theory (New York: Academic Press)
3. Chang M.G., Hatzakis M., Wilson A.D. and Broers A.N. (1977) "Electron beam lithography draws a finer line" Electronics (12 May) 89-98
4. Keyes R.W. (1975) "Physical limits in digital electronics" Proc. IEEE 63 (May) 740-767
5. Murata K., Matsukawa T. and Shimizu R. (1971) Japan J. Appl. Phys. 10, 673-686
6. Wallmark J.T. (1979) IEEE Trans. Electron Devices ED-26, 135-142

Springer Series in Solid-State Sciences

Editors: M. Cardona, P. Fulde, H.-J. Queisser

Springer-Verlag
Berlin Heidelberg New York

Selected Iss

Lecture Notes in Mathematics

Vol. 662: Akin, The Metric Theory of Banach Manifolds. XIX, 306 pages. 1978.

Vol. 665: Journées d'Analyse Non Linéaire. Proceedings, 1977. Edité par P. Bénilan et J. Robert. VIII, 256 pages. 1978.

Vol. 667: J. Gilewicz, Approximants de Padé. XIV, 511 pages. 1978.

Vol. 668: The Structure of Attractors in Dynamical Systems. Proceedings, 1977. Edited by J. C. Martin, N. G. Markley and W. Perrizo. VI, 264 pages. 1978.

Vol. 675: J. Galambos and S. Kotz, Characterizations of Probability Distributions. VIII, 169 pages. 1978.

Vol. 676: Differential Geometrical Methods in Mathematical Physics II, Proceedings, 1977. Edited by K. Bleuler, H. R. Petry and A. Reetz. VI, 626 pages. 1978.

Vol. 678: D. Dacunha-Castelle, H. Heyer et B. Roynette. Ecole d'Eté de Probabilités de Saint-Flour. VII-1977. Edité par P. L. Hennequin. IX, 379 pages. 1978.

Vol. 679: Numerical Treatment of Differential Equations in Applications, Proceedings, 1977. Edited by R. Ansorge and W. Törnig. IX, 163 pages. 1978.

Vol. 681: Séminaire de Théorie du Potentiel Paris, No. 3, Directeurs: M. Brelot, G. Choquet et J. Deny. Rédacteurs: F. Hirsch et G. Mokobodzki. VII, 294 pages. 1978.

Vol. 682: G. D. James, The Representation Theory of the Symmetric Groups. V, 156 pages. 1978.

Vol. 684: E. E. Rosinger, Distributions and Nonlinear Partial Differential Equations. XI, 146 pages. 1978.

Vol. 690: W. J. J. Rey, Robust Statistical Methods. VI. 128 pages. 1978.

Vol. 691: G. Viennot, Algèbres de Lie Libres et Monoïdes Libres. III, 124 pages. 1978.

Vol. 693: Hilbert Space Operators, Proceedings, 1977. Edited by J. M. Bachar Jr. and D. W. Hadwin. VIII, 184 pages. 1978.

Vol. 696: P. J. Feinsilver, Special Functions, Probability Semigroups, and Hamiltonian Flows. VI, 112 pages. 1978.

Vol. 702: Yuri N. Bibikov, Local Theory of Nonlinear Analytic Ordinary Differential Equations. IX, 147 pages. 1979.

Vol. 704: Computing Methods in Applied Sciences and Engineering, 1977, I. Proceedings, 1977. Edited by R. Glowinski and J. L. Lions. VI, 391 pages. 1979.

Vol. 710: Séminaire Bourbaki vol. 1977/78, Exposés 507–524. IV, 328 pages. 1979.

Vol. 711: Asymptotic Analysis. Edited by F. Verhulst. V, 240 pages. 1979.

Vol. 712: Equations Différentielles et Systèmes de Pfaff dans le Champ Complexe. Edité par R. Gérard et J.-P. Ramis. V, 364 pages. 1979.

Vol. 716: M. A. Scheunert, The Theory of Lie Superalgebras. X, 271 pages. 1979.

Vol. 720: E. Dubinsky, The Structure of Nuclear Fréchet Spaces. V, 187 pages. 1979.

Vol. 724: D. Griffeath, Additive and Cancellative Interacting Particle Systems. V, 108 pages. 1979.

Vol. 725: Algèbres d'Opérateurs. Proceedings, 1978. Edité par P. de la Harpe. VII, 309 pages. 1979.

Vol. 726: Y.-C. Wong, Schwartz Spaces, Nuclear Spaces and Tensor Products. VI, 418 pages. 1979.

Vol. 727: Y. Saito, Spectral Representations for Schrödinger Operators With Long-Range Potentials. V, 149 pages. 1979.

Vol. 728: Non-Commutative Harmonic Analysis. Proceedings, 1978. Edited by J. Carmona and M. Vergne. V, 244 pages. 1979.

Vol. 729: Ergodic Theory. Proceedings 1978. Edited by M. Denker and K. Jacobs. XII, 209 pages. 1979.

Vol. 730: Functional Differential Equations and Approximation of Fixed Points. Proceedings, 1978. Edited by H.-O. Peitgen and H.-O. Walther. XV, 503 pages. 1979.

Vol. 731: Y. Nakagami and M. Takesaki, Duality for Crossed Products of von Neumann Algebras. IX, 139 pages. 1979.

Vol. 733: F. Bloom, Modern Differential Geometric Techniques in the Theory of Continuous Distributions of Dislocations. XII, 206 pages. 1979.

Vol. 735: B. Aupetit, Propriétés Spectrales des Algèbres de Banach. XII, 192 pages. 1979.

Vol. 738: P. E. Conner, Differentiable Periodic Maps. 2nd edition, IV, 181 pages. 1979.

Vol. 742: K. Clancey, Seminormal Operators. VII, 125 pages. 1979.

Vol. 755: Global Analysis. Proceedings, 1978. Edited by M. Grmela and J. E. Marsden. VII, 377 pages. 1979.

Vol. 756: H. O. Cordes, Elliptic Pseudo-Differential Operators – An Abstract Theory. IX, 331 pages. 1979.

Vol. 760: H.-O. Georgii, Canonical Gibbs Measures. VIII, 190 pages. 1979.

Vol. 762: D. H. Sattinger, Group Theoretic Methods in Bifurcation Theory. V, 241 pages. 1979.

Vol. 765: Padé Approximation and its Applications. Proceedings, 1979. Edited by L. Wuytack. VI, 392 pages. 1979.

Vol. 766: T. tom Dieck, Transformation Groups and Representation Theory. VIII, 309 pages. 1979.

Vol. 771: Approximation Methods for Navier-Stokes Problems. Proceedings, 1979. Edited by R. Rautmann. XVI, 581 pages. 1980.

Vol. 773: Numerical Analysis. Proceedings, 1979. Edited by G. A. Watson. X, 184 pages. 1980.

Vol. 775: Geometric Methods in Mathematical Physics. Proceedings, 1979. Edited by G. Kaiser and J. E. Marsden. VII, 257 pages. 1980.